THE POWER OF THREE

Thomas Fowler
Devon's Forgotten Genius

Pamela Vass

With additional material by
Mark Glusker and David Hogan

Pamela Vass.

www.boundstonebooks.co.uk

First published in Great Britain in 2016 by Boundstone Books, Little
Boundstone, Littleham, Bideford. EX39 5HW.

ISBN: 978-0-9568709-5-7

Cover image: Great Torrington High St and Market
Square. Print from an engraving of 1850 by Kearshaw & Son
Maths detail © Ffatserifade Dreamstime.com

Printed and bound in Great Britain
by SRP Exeter

www.boundstonebooks.co.uk

Also by Pamela Vass

Fiction
Seeds of Doubt
Shadow Child

Non-Fiction
On Course for Recovery

Edited Works
In My Own Words

Acknowledgements

My thanks for their support of, and interest in, this work go to:
David Avery, David Batty, Nicholas Bodley, Professor Nikolay Brousentsov, The Devonshire Association, Peter Christie, Pip Jollands, Great Torrington and District Community Development Trust, Great Torrington Museum, Alison Harding, Roy Foster, Professor Ralph Merkle, North Devon College, Dr. Doron Swade MBE, Professor Gary Tee and Torridge Leader Company.

For allowing access to, and use of, their archive material to:
British Library, British Museum, Bodleian Library Oxford, Cambridge University Library Royal Greenwich Observatory Archive, Lord and Lady Clinton, Devon Record Office, Devon & Exeter Institution, Gloucestershire Archives, Great Torrington Museum, Institute of Civil Engineers, King's College London, London Metropolitan Archive, North Devon Athenaeum, Radcliffe Science Library, Royal Astronomical Society, Royal Society, Science Museum, Trinity College Cambridge, Totnes Museum, University College London, University of London Library, Westcountry Studies Library, Wren Library Cambridge.

To Roger Carter, John Dingle, Patrick Kivlin, Sue Scrutton, Laurence Shelly, Terence Sackett and Michelle Taborn for their thorough scrutiny and invaluable feedback.

To John McKay for his generosity in sharing his knowledge of Thomas Fowler.

My particular thanks go to Mark Glusker for the commitment and skill he has brought to the interpretation of Fowler's work and for his many other contributions.

Also to David Hogan, for his conviction that the full potential of Fowler's work has yet to be realised and for his many contributions to this research.

Introduction

"WOODEN COMPUTER INVENTED IN NORTH DEVON". A striking headline for a provincial newspaper. The article in the *North Devon Journal* continued "It is fascinating ... to know that one of the original pioneers of the computer was a self-taught bookseller and printer of Torrington who was born over 200 years ago. His name was Thomas Fowler." I was intrigued, especially given that Great Torrington is close to where I live. But surely 'the father of computing' was Charles Babbage? Who was Thomas Fowler?

It was a question that was to take me on an emotional journey, from excitement at this charismatic inventor's early success to despair at his betrayal, from admiration of his ingenuity to the agony of obscurity. Despite an initial reluctance to peer over the precipice into the world of mathematics and the history of computing, I knew this was a story that had to be told.

My first searches revealed nothing. Then I met local businessman, John McKay, who had discovered a brief biography of Fowler. It depicted an extraordinary, self-taught mathematician and inventor who transformed himself from fellmonger to successful businessman and pillar of the community. A significant achievement in itself; but more was to come — much more. Thomas Fowler invented a unique calculating machine. Not something that would normally inspire me, but this was different. His absolute belief that this invention would be his claim to fame leapt off the page. So why had I never heard of him? Had his invention survived? If not, did any drawings exist? Could it be recreated?

Curiosity and a nagging need for justice launched me on a search for clues, most of them buried deep in the archives of the most respected scientists of the nineteenth century. Along the way I discovered an unexpected fascination with the leaps of imagination that lie behind the birth of the modern computer. All leading to one startling conclusion.

The computer is an indispensable part of daily life. We rely on it to communicate, for entertainment and for access to a wealth of resources that our bookshelves could never hold. Yet many of us are ignorant of the processes that make all this possible — with one exception. It is well-known that the technology relies on the binary number system, a choice made by early computer scientists that dominates the world of computing. But a fundamentally different path might have been taken more than a century earlier had the pioneering work of this humble, self-taught mathematician been adopted.

For a while there was every chance that it would. Yet after a flurry of recognition, Fowler's work was consigned to obscurity. Why?

This book tells the fascinating story of a remarkable Devon man who was convinced that his innovative work was overlooked because of his humble origins and unorthodox path to mathematical brilliance. It includes a layman's view of Fowler's inventions and a full appendix containing his detailed mathematical deliberations and original source material for those who would like to explore further.

Thomas Fowler died in 1843 but the final chapter of his story has yet to be written. As twenty-first century scientists rediscover his ground-breaking work, perhaps there is still time for Fowler's dying wish to be fulfilled.

"My greatest wish was to have had a thorough investigation of the whole principle of the Machine ... by some first rate Man of Science before it be laid aside or adopted."

A Memoir

In which I anticipate the first demonstration
of my unique invention.[1]

Could I have done things differently? Perhaps said a little more, or a little less? Been wiser in the ways of men? Had I known what lay before me I might have reconsidered the path I had chosen to bring my unique calculating machine to public notice. But, being ignorant of the prejudice that lay at the very heart of this country's scientific community, I did not.

Thus, as dawn broke on the morning of May 13th 1840, I readied myself for the journey to London. My dear wife Mary had been calling for some time, anxious that the coach should not leave without me. But it was imperative that I personally supervise the transport of the mechanism. I had risen before dawn to pack and re-pack the frame, cushioning all the moving parts to ensure they survived the journey unscathed. This was the price I had to pay for living and working so far from those esteemed — as I then thought — persons who formed this country's scientific elite.

It was not easy, manoeuvring the wooden container between the presses of my print shop. It was all of three feet by four by one and took three of us to carry it safely to the cart. These dimensions were far larger than my intended machine but I was constrained by the tolerances that working in wood afforded me. I had every expectation that, having proved the worth of my invention, I should be able to demonstrate a mechanism barely half that size in time. I tremble now at what was actually to transpire.

But I am ahead of myself. On that Wednesday morning, as I stepped into the square in Great Torrington, I was fully confident of acclaim for an invention, which I did then, and do now, in all modesty, believe to be my greatest claim to fame.

---

[1] Inspired by Thomas Fowler, imagined by Pamela Vass

Chapter 1
[+]

The Reluctant Apprentice

High Street & Market Place.

On Wednesday May 13th 1840, Thomas Fowler turned his back on a peaceful retirement and travelled to London to keep the most important appointment of his life at 47 Hunter Street, Brunswick Square. He took with him a deceptively simple hand-built device; a device with the power to turn the world of calculation upside down.

It was a formidable undertaking for a sixty-two year old man who had spent his entire life in the small market town of Great Torrington in the west of England. He was born on June 9th 1777, in a typical cob and thatch cottage bordering a narrow street in the heart of the working quarter of town. Beside the cottage a dark alleyway opened out into a yard surrounded by rough wooden buildings where his father, Hugh Fowler, carried on his trade as a cooper. Hugh had moved to Great Torrington from nearby Broadwood Kelly a short time before his marriage to Elizabeth Scading on July 10th, 1775. As the eldest son, Thomas learnt the coopering trade alongside his father, helping him meet the steady demand for casks from the twenty or more inns in the town. For many boys that would have been the extent of their education, but Thomas was lucky. He joined the twenty-one other pupils at the Bluecoat Charity School, held in the building next door to this home in Well Street.

The school provided little more than a basic education during hours snatched out of the working day, but it gave the young Thomas a thirst for learning. He soon mastered reading and writing and, more significantly, discovered a flair for arithmetic. There was no disguising his ability. At the age of thirteen he was awarded a £1 premium towards an apprenticeship, regarded as the best start in life by his peers. But Thomas probably did not see it that way. Having glimpsed a world where all things were possible with an education, he would have dreamt of going on to grammar school, or even university.

Instead, he faced the harsh reality of the fellmonger's yard. The smell leeching into the street was enough to turn anyone's stomach and once inside, the seemingly endless days were full of hard, physical labour. He worked from dawn to dusk stripping fat and hair from the animal hides that arrived fresh from the abattoir before being passed on to the tanner's yard for processing. At the end of a day the stench of the yard must have lingered in his skin, hair and trademark leather apron.

It is difficult to imagine a worse fate for a boy with such potential. He would have adjusted to the sights and smells of the yard over time, but shaking off a growing discontent with his life would have been much harder. There was only one thing keeping his dream of a different future alive, a book — for a long time the only one he possessed — John Ward's *Young Mathematician's Guide*. It is impossible to know for sure how he came by it but we have a clue in a subscription list issued by London publishers, John Hellins. In 1788, Torrington cleric the Reverend John Poole subscribed to *Mathematical Essays on several subjects containing new improvements and discoveries in the mathematics*. Given the Bluecoat School's strong connections with the church perhaps this was the man who inspired Fowler and gave him his copy of the *Young Mathematician's Guide*.

What is certain is that Fowler was determined to pursue his fascination with mathematics. While other workers at the yard made their way to the Setting Sun Inn at the end of the day, he returned to his room, took down a tallow candle, and began to study. His son, Hugh, later said of his father:

> This book [*Ward's*], as is usually the case with the *homo unius libri*, he thoroughly mastered, and that without the slightest help from any one. No one could have been more entirely self-taught than he was. Mathematicians in those days were very scarce in this part of Devonshire, and probably elsewhere, even in the great centres of education … Few people knew, or if they knew cared, that there was in their midst "a wondrous boy", who absolutely self-taught, after his hard day's work among sheepskins spent half the night poring over his mathematics.[2]

This leather-bound volume transformed Fowler's life, introducing him to a world he could navigate almost by instinct. It was as though the answers were there, waiting to be discovered — he simply had to frame the right questions. The satisfying beauty and balance to mathematical equations, the sense of everything falling into place through the adoption of symbols that meant more than the written word, obviously struck a chord. He had found his passion, one that would remain with him until his dying day.

Fowler was bound as an apprentice for seven years, but even then, without capital or private means his choices were limited. For a while, life in Great Torrington was also overshadowed by the French Revolution. Shock waves reached across the channel causing the Prime Minister, Pitt the Younger, to recruit a new body, the Volunteers for Home Defence. He appealed for men to join "a species of Cavalry consisting of gentlemen and yeomanry who would not be called upon to act out of their own counties except under pressure of invasion or urgent necessity." Thomas Fowler's life took on a very different colour when he signed up. Once a week he turned out on horseback for drill, resplendent in his scarlet jacket with yellow facings, silver lace and white breeches, and wielding a pistol and sword. These drills, along

---

[2] See appendix 1

7

with fourteen days permanent duty in quarters each year, gradually shaped the Volunteers into some kind of fighting force.

But time passed, the threatened invasion failed to materialise, and Fowler was able to re-focus on his own future. In 1805 an advertisement appeared in the *Exeter Flying Post*: "Printing Types, Press, Bookbinding Presses; Bookseller — stock in trade, and a great variety of bookbinding tools, mostly new." Torrington's only printer, Mr. Wilson, was selling up. With his passion for study Fowler would have been well acquainted with Mr. Wilson, but a mutual interest in books may well have become something much more significant — his key to a new life.

Within a few years Fowler had established himself as a printer and bookseller at a property in the High Street. These were exciting times for a young man still in his twenties. Not only had he escaped the fellmonger's yard but he was now the boss of his own business. And he was about to flex his creative muscles for the first time. His son, Hugh, later recorded "his printing-machine, by the way, he made with his own hands on a plan of his own invention." Combining the woodworking skills he had learnt from his father in the cooper's yard with a natural creativity, Fowler demonstrated an ability that was to become his hallmark — a flair for devising simple solutions to complex problems, a talent for invention.

From his print shop, Fowler produced and distributed everything from local pamphlets to council papers, publican's paperwork to posters of all kinds. He also stocked books for those who could afford this luxury, and traded as a bookbinder, attracting the patronage of the landed gentry happy to avoid sending precious volumes by carrier to London. One particularly fine book, *Flora Ceres and Pomona*, written by John Rea in 1665, was re-bound by Fowler in 1834 for Lord Rolle's library at his house at Stevenstone, just a mile or so to the east of Torrington.[3]

Building his business took time. Many in Great Torrington were suffering from heavy taxation and food shortages and forced to seek public help under the old Poor Law. The price of bread soared as demand exceeded supply. For those living at subsistence level it was

---

[3] This volume is on loan to the Thomas Fowler exhibit at Torrington Museum.

the final straw and protests erupted across the county. In these pre-police force days tackling disturbances was a thorny problem. But the government now controlled the Yeomanry — what better way to suppress riot and rebellion? Fowler may well have found his newly acquired military skills employed not against a foreign enemy but in maintaining order at home.

When not on Yeomanry duty, all the evidence is that Fowler was focused on one thing; securing his place as a businessman within Great Torrington. He spent more than half his lifetime in single-minded pursuit of this goal — until he met Mary Copp.

Mary was also Torrington born, the second daughter of Charles and Christian Copp, a well-known family of carriers within the town. With a range of carts and wagons they transported anything from livestock to furniture about the county. She was considerably younger than her husband-to-be; on February 21st, 1813, their wedding day, he was thirty-five, she was just twenty-one. But when Mary married Fowler she became his partner in all senses of the word.

This exceptional woman was to play a significant part in his professional as well as his personal life. Both Fowler and Mary came from sizeable families, with three sisters and three brothers apiece, but over the next twenty-three years Mary became mother to their eleven children. The war was still rumbling on when their first child, Caroline, was born in the spring of 1814. Two years later Hugh arrived, followed by Mary in 1817, and then a new arrival every two years as Thomas was followed by Cecilia, Henry, Charles and Frances Honor. Finally, with three-year gaps this time, Newell Vicary and the twins, Paul and Silas.[4]

Caroline quickly became invaluable in the business. In 1822 local man, Daniel Johnson returned home to Torrington after serving as a surgeon in the East India Company. He asked Fowler to publish a manuscript of his experiences and Caroline, although only eight years old, worked alongside her father on the commission. The concluding paragraph of Johnson's book records "the greatest part of this book was composed by a child of not more than eight years of age, Caroline Fowler, a daughter of the printer." Imagine the intelligence cascading through this family when a child this young was literate enough to

---

[4] See appendix 15 for the Fowler family tree

master putting individual letters back to front in a wooden frame ready
for printing.

SKETCHES
OF
FIELD SPORTS
AS FOLLOWED BY
The Natives of India
WITH OBSERVATIONS ON THE ANIMALS.
ALSO

An account of some of the customs of the Inhabitants,
and natural productions, Interspersed with various
Anecdotes.

LIKEWISE THE LATE NAWAB VIZIER ASOPH UL DOW-
LAH'S GRAND STYLE OF SPORTING AND CHARACTER.

A DESCRIPTION OF THE
ART OF CATCHING SERPENTS,

As practised by people in India, known by the appellation of
Conjoors, and their method of curing themselves when bitten.

WITH REMARKS ON
Hydrophobia, & Rabid Animals.

By DANIEL JOHNSON,

FORMERLY SURGEON IN THE HONORABLE EAST INDIA COM-
PANY'S SERVICE AND RESIDENT MANY YEARS AT CHITTRAH IN
RAMGHUR.

Utilissimum sæpè quod contemnitur. Phæd.

LONDON:
PUBLISHED FOR THE AUTHOR, BY LONGMAN, HURST,
REES, ORME, AND BROWNE, AND THOMAS FOWLER.
GREAT TORRINGTON, DEVON.

1822.

Caroline became quite a celebrity, with the *Bristol Mirror* writing of a *Great Literary Curiosity*.

> In announcing the publication of *Indian Field Sports*, a London Literary Journal gives the following instance of precocity of talent exerted in producing the work: – The printing was almost entirely performed by a girl under nine years of age (the Clara Fisher of typography) at a press made by her father, Mr. Fowler of Torrington, Devon, of which press and infant compositor the *Indian Field Sports* are the first fruits.

By 1825, Fowler had seven children under eleven, an exhausting reminder of the need to look for additional ways to provide for his family. The solution lay in the local bank, Loveband, Slade & Co. For ten years he acted as clerk and scrivener before his obvious mathematical abilities led to promotion to assistant manager, a significant accolade for a self-made man.

Fowler was now well on the way to the future he envisaged for himself as a young boy. But every step he took was the direct result of his own initiative, ability and dogged determination. Without private means, pursuing his studies remained an uphill battle, but he refused to allow this to deter him. With Mary's support he eagerly grasped every opportunity to combine his business responsibilities with his passion for mathematics and problem solving.

Chapter 2
[+ −]

In which Fowler is told to go hang
or drown himself

"Wooden Computer invented in North Devon" was the headline that launched me on the track of Thomas Fowler, but it was information about another of his inventions that came my way first.

In 1826, alongside his work in the bank and print shop, Fowler became preoccupied with the problem of heating his glasshouse. Dissatisfied with existing methods, he looked for a better solution. He rejected the simple open fireplace; brick flues gave out a sulphurous gas that damaged the plants and even using two fireplaces it was impossible to keep an even temperature. Looking for alternatives, Fowler may have been inspired by two Frenchmen, Bonnemain and Chabbannes. Bonnemain experimented with heating by circulating hot water to hatch chickens, then moved from incubators to greenhouses. But it was the Marquess de Chabbannes, who applied this new method of warming using hot water circulation at his factory in London.

James Watt, famous for his steam engine, also used steam and hot water for heating rooms, baths and for manufacturing. But a steam boiler was expensive and needed frequent repairs, the fire still required constant attention and there was an alarming risk of explosions at the hands of inexperienced garden assistants. Hot-water systems were more promising, like one being developed by William Atkinson, a Chelsea architect. One of his installations delighted the gardener who could at last get a good night's sleep.

But Fowler was convinced the circulation of the water could be improved. After devoting, as he later said, "long hours in solitude and silence", to his experiments, he made an intellectual leap that had escaped everyone else. The principle of convection — where hot water rises and cold water falls to replace it — was at the heart of every sealed hot water system. But artificial pressure was needed to circulate the water. Then Fowler had a revelation. The secret to dispensing with artificial pressure was right in front of him, and above him — in fact all around him. The answer lay in employing atmospheric pressure.

To prove his point, Fowler designed a simple apparatus to "elevate and circulate hot fluid from an open boiler or vessel, containing the fluid, without the external application of any mechanical force or pressure whatsoever, except the common pressure of the atmosphere. I propose to call this instrument, under all its different modifications, a THERMO – SIPHON." In its simplest form the Thermosiphon is a bent tube with the ends immersed half way down water held in two vessels joined by a connecting tube at the bottom.

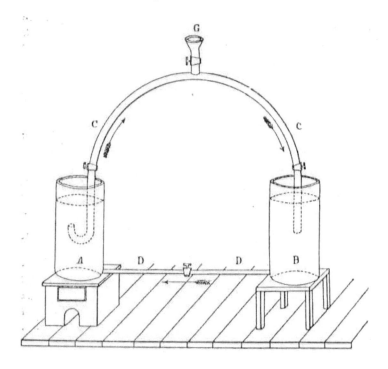

Fowler installed a Thermosiphon in his garden:

It is really astonishing to see the effect of the Thermosiphon in causing a circulation of boiling water through this case (a lead case, three feet wide by four long, and two and a half inches deep laying a foot below the surface) Its action is perfectly tranquil and sure, without the slightest danger; and the soil above the case gets warm very fast, although surrounded by the freezing atmosphere of the last three or four very cold nights. If the earth be turned up it is found smoking hot. This induced me, last Saturday night, in a very hard frost, to sow some mustard, radish, and other seeds, on the surface of this small hot-bed, which I covered with a thin deal box, and now (Monday afternoon) I find the seeds are germinating; although the whole is subject to the keen blast of one of the coldest days we have had here for a long time.[5]

Torrington residents know how keen those blasts can be! Fowler's success prompted him to approach John Sloley, a Gentleman with a large vinery behind his home at 28 South Street, Torrington, just a few yards from his Printing Office. It became quite a talking point. When Mr. Sloley decided to rent his house, the advertisement specifically mentioned that "the hot house and greenhouse are heated on the principle of Mr. Fowler's Patent Thermosiphon." Word soon spread, with Lord Rolle commissioning Fowler to install his system in the glasshouse on his estate at Bicton near Exeter. *The Exeter and Plymouth Gazette* picked up on this:

A few weeks ago we took occasion to notice a philosophical apparatus or machine for communicating heat termed a Thermosiphon, invented by Mr. Fowler of Torrington. We are informed that one of these extraordinary machines, on a scale of the most magnificent description, is now completing at Bicton for the Right Hon Lord Rolle; and that two more on a very extensive scale are erecting in the immediate vicinity of Exeter under the direction of Mr. Coldridge, ironmonger. This

---

[5] Ref: Institute of Civil Engineers Archives. Tracts 8vo Vol 58. Fowler's Thermosiphon

invention has been highly spoken of in *The Gardener's Magazine* and it is but justice to Mr. Fowler to say that he has made a discovery which has escaped the research of every philosopher and scientific engineer up to the present day.

"It is but justice to Mr. Fowler to say that he has made a discovery which has escaped the research of every philosopher and scientific engineer up to the present day." Quite an accolade. Working completely alone, isolated from others exploring this problem, Fowler had come up with the solution. And he was already marketing his invention. Mr. Coldridge advertised "Fowler's celebrated Patent Thermosiphon or Hot Water Machine for heating Conservatories, Vineries, Mansion House, Churches, Hospitals and Asylums." One of the two erected by Mr. Coldridge was at Coaver, the residence of J. Milford, Esq. an Exeter banker. Henry Dalgleish, his head gardener, sent a glowing review to *The Gardener's Magazine*. "I have seen several steam-machines, level hot water apparatus, and many different constructions of brick flues, for causing artificial heat, but am decidedly of opinion that the Thermosiphon is very superior to all other plans hitherto adopted."

This was a major achievement for Fowler. His first project, the printing press, was a significant step but hardly headline material. This time his invention had been picked up, not only by local newspapers, but also nationally. *The Gardener's Magazine* carried a review:

Mr. Fowler has had the good fortune to hit on the idea that water may be heated and made to circulate through a siphon, as well as through horizontal pipes, or by force through pipes in any direction; provided always, that the height of the siphon be not greater than to be counter-balanced by the pressure of the atmosphere, say not greater than 30ft. Any person might have discovered the same thing by reflection, or in answer to the question asked; but we are not aware that the idea has occurred to ... any of the numerous engineers now occupied in applying this mode of heating. Mr. Fowler's discovery is not likely to be of very important use in gardening ... but it will be of most advantageous application in private houses for heating baths, apartments, water for washing, etc[6].

The editors became carried away with other possibilities, particularly heating baths. A distinct advantage was that hot water might be "let into the bath at pleasure without troubling servants, or indeed, without their knowing anything of what is going on."

Fowler listed other applications, including heating a hot plate for copper-plate printing. He also wrote, rather prophetically "I can with confidence say, that in the hands of any skilful person acquainted with the principles of hydraulics and pneumatics, it [the Thermosiphon] may be carried to a very great extent, and the forms of its apparatus may be varied, according to circumstances, almost infinitely, both for heating houses and buildings, and for all horticultural purposes requiring artificial heat."

Fowler was probably delighted by *The Gardener's Magazine* review, if irritated by the words "has had the good fortune to hit on the idea." The phrase invokes a vision of inventions wafting around in the ether simply waiting to be reeled in by anyone who chances on them. Far from being a stroke of good fortune, Fowler's breakthrough was the result, as he said, of long hours in solitude and silence. Significantly, it was a discovery he made ahead of the most respected inventors of his generation. Not only this, but the fundamental principle is still in use today, not in horticulture, but, as *The Gardener's Magazine* picked up, for heating our homes.

Given this achievement, why is Thomas Fowler not a household name?

It was not for want of trying.

After the review in *The Gardener's Magazine* brought his pioneering work to the attention of the whole country, the obvious next step was to apply for a Patent. Clowes & Co, the London firm handling Patents, advised Fowler to draft the specification from which a solicitor would prepare the final document. Fowler selected Henry Brougham, a well-respected lawyer and social reformer. However, his local solicitor

---

[6] *Gardener's Magazine* references ©British Library ST 1894

directed him to a Mr. B. Rotch of Furnival's Inn, advice he reluctantly accepted and later bitterly regretted.

Towards the end of November 1828, Fowler was summoned to London by Mr. Rotch to look over the specification. He was unhappy with some of the wording and so, the next morning, went to Clowes & Co to change it. Unfortunately the document had already been completed and simply needed his signature. An interesting then situation developed. Fowler was determined to add another useful application of the invention but the officers of Clowes and Co, with their considerable experience in handling Patent documents, attempted to dissuade him. "This [making additions to the document] I was informed seemed to offend Mr. Rotch but I was determined to act according to my judgement, honestly & fairly by describing what I did then and do now consider a most useful and extensive application of the invention which I am certain will hereafter give much employment and prove highly beneficial."

A principle was at stake and once set on a course of action Fowler was not easily diverted. After further unsuccessful protests the staff agreed to compromise by adding another sheet to the document, leaving Mr. Rotch's piece untouched. A solution of sorts had been arrived at and the matter was closed. Or so Fowler thought.

Sadly the consequences of his actions, however well founded, were to haunt him for the rest of his life.

Fowler delivered the amended specification to the Chancery office and returned home to Torrington thinking all was safe. A few weeks later this feeling was reinforced when he opened a letter from the Patent Office confirming he had been successful in obtaining Patent no. 5711, *Specification of Thomas Fowler; an Apparatus for Raising and Circulating Hot Water , Etc.* The Patent secured, Fowler should have been assured of a modest return for his inspirational work, if not the place in history his discovery warranted.

A

DESCRIPTION

OF THE

PATENT THERMOSIPHON;

WITH

SOME MODES OF APPLYING IT

TO

HORTICULTURAL,

AND OTHER

USEFUL AND IMPORTANT PURPOSES.

Illustrated with Plates.

INVENTED BY

.THOMAS FOWLER,

GREAT TORRINGTON, DEVONSHIRE.

LONDON:

Printed for the Patentee,

AND SOLD BY LONGMAN, REES, ORME, BROWN, AND GREEN,

PATERNOSTER-ROW.

1829.

18

But Fowler's satisfaction was short lived. Just a few months later he received some shattering news. Cottam & Hallen and other firms in London were openly infringing his Patent. Mr. Cottam had apparently also 'discovered' the effect of atmospheric pressure, coincidentally just a month after Fowler lodged his Patent. Fowler immediately wrote to Mr. Rotch for advice and received an extraordinary reply. The solicitor informed him that, because of the alterations he had insisted on making to the Patent document, his claim could not be supported in any Court of Law. In another letter, Fowler claims Rotch then said "that I had no other resource but to hang or drown myself according to any particular inclination I might have for either of those respectable modes of getting rid of all my troubles." An extraordinary response!

Was this malicious retort because Mr. Rotch was so deeply offended by his client's rejection of his professional advice that he took satisfaction in his downfall? Perhaps Fowler's single mindedness came across as arrogance, and Mr. Rotch thought he deserved his comeuppance. But there was more. Conveying the bad news was one thing, but engineering it was quite another. This was certainly Fowler's impression. "In my perhaps not singular case I am sacrificed for being too honest, but the most annoying circumstance is, that I am compelled to look on Mr. Rotch as a principal Agent in this affair as I am credibly informed that Cottam & Hallen applied to him for an opinion and that he told them they might infringe my patent with impunity." Someone paid to act on his behalf had blatantly sided with his competitors.

The cost, both financial and personal was shattering According to Fowler, he had ruined himself with the expense of the Patent and his experiments and could not afford to take Cottam and Hallen to court. He was left without any outlet for his feelings of injustice and betrayal, feelings compounded by later events. In 1837, Charles Hood published a pamphlet on the development of hot water heating systems but claimed he could find nothing except a few scattered and unimportant notices on the subject. Was Hood aware of Fowler's Patent, lodged eight years earlier, but decided not to credit him? Or was it pure co-incidence that their work followed similar directions? Many Patents lodged during the 1830s were for successive improvements on existing procedures or devices and crediting original work could be difficult.

A pamphlet published in France in 1844, and actually entitled: *Application of the Art of Heating by the Thermosiphon*, credits Bonnemain, Chabbanes, Bacon, Atkinson and Charles Hood, but makes no reference to Fowler. The author simply comments (translated from the French), "At this time [1837] the thermosiphon, abandoned for some time in France, became fashionable." Sadly there are no clues as to whether "some time" refers to pre or post Fowler.

With his ground-breaking invention either pirated or overlooked on all sides, life was becoming a constant struggle against the odds for Fowler. And more was to come. In June 1832 Fowler's second daughter, Mary, died, aged just fourteen. But sadly Fowler and Mary's grieving was far from over. Only three years later they lost their second son, Thomas, aged just sixteen. This time of personal tragedy undoubtedly fed Fowler's feelings of anger and bitterness. Over five years on from his Patent dispute, Fowler felt compelled to write to Lord Brougham, now Lord Chancellor:

> My Lord, As the time appointed for the commencement of the next Sessions of Parliament is fast approaching your Lordships public virtue and great talents will again be called in to action, and I truly believe, for the benefit of the human race among other matters the Laws of Patents for inventions will in all probability be again taken in to consideration and such amendments be made as may in future preserve the poor ingenious Man from almost certain ruin as the only reward of his ingenuity and as every species of information may be useful at the present moment I humbly and respectfully beg to lay before your Lordship my case as a Patentee, a case which indeed sends me back angry and disappointed to an "almost starving family".

Fowler relates the facts and Mr. Rotch's part in them. If anything the intervening years had only intensified his anger:

> It is very hard that any poor man that ventures to take out a Patent for what he considers a useful invention and spends his all in obtaining it should be compelled to give up all the benefit originally intended by the legislature; or, immediately after

become the Prey of the Caste of needy and too often rapacious Lawyers.

Mr. Rotch had certainly left his mark. Fowler concluded that some men might indeed be driven to drown or hang themselves if they continued to be treated in this way by the present Patent Laws. There was often no other path open to them. As for himself, he continued:

> I do not even now believe that my specification vitiates the Patent but I have no means of bringing the matter before your Lordship in a legal form & must therefore content myself with the loss although poverty is absolutely staring me in the face in consequence of the Patent Laws and of the power the English Lawyers now have of making the worse appear the better reason; this Patent with expenses has cost me above £400 most of which I was obliged to borrow so that I may say I am ruined.[7]

Fowler had hoped to reap a reward for the years spent developing his Thermosiphon. To achieve this, he not only handed over all his plans, but also insisted on altering the Patent document to ensure it was completely accurate — with devastating consequences. But was he defeated by bureaucracy or by his own obstinacy? He was certainly not a man easily deflected from the truth as he saw it, even in the face of contrary professional advice.

A more sympathetic solicitor might have recognised and accommodated Fowler's quirk of character. Instead Rotch reacted angrily to his inexperienced client, this unworldly countryman who thought he knew better than the professionals and who deserved his comeuppance. Whatever the truth of it, Fowler was thrown into turmoil by this perceived injustice. Feelings of betrayal became trapped deep within his psyche, affecting his thoughts and actions not only in the wake of this confrontation, but for the rest of his life.

---

[7] Ref: UCL Library Services, Special Collections. See Appendix 14

Chapter 3
[+ +]

As Easy as One, Two, Three

After all the hope and promise of the past few years, the rug had been pulled from beneath Fowler. If only he had not applied for a Patent; if only he had not engaged Mr. Rotch; if only he had had the means to challenge him in court. Everything he had worked so hard to build was now at risk. He had nine children at home. How was he going to support them with a £400 loan to repay? He was relatively restrained in his letter to Lord Brougham, but the anger and bitterness are clear. Fowler's response was to bury himself in his work, increasing his load by taking on two more posts. How he found the time is a mystery. Fortunately, his daughters, Caroline and Cecilia, were now working in the family business but even with their help, finding space for new responsibilities must have been a challenge.

As well as his formal duties in the Printshop and Bank, Fowler took on the challenge of becoming a town councillor. A new broom was desperately needed. A Commissioner's report of 1834 had highlighted the poor government of the town. They singled out the administration of justice as being particularly defective "owing to the want of attention and vigour on the part of the magistrates, who are afraid of punishing offenders. There is consequently much degradation in the town, which is very disorderly, and the offenders, though discovered, often escape without punishment."

Not everyone got away with it though. In 1828 a debtor's body was disinterred from Torrington graveyard and conveyed to Exeter prison to serve out his sentence!

Fowler was one of those newly appointed to the council with the specific task of drafting "Bye Laws for the good government of this Borough". All voluntary of course, but there were side benefits. At the bottom of the new Bye Laws and other council stationery are the words "Printed by T. Fowler".

He was a conscientious attender of meetings, determined to act as he thought right and just. When Lord Rolle tried to charge Torrington residents for taking earth and stone from the Commons he supported the Council's expression of "surprise and regret that powerful individuals should attempt to wrest from the inhabitants a privilege which is important and valuable to the poor". Whether the decision was on the introduction of gas to the town, lockups under the town hall, lanterns for the night constables or the new Market House, Thomas Fowler was dutifully in his place ready to vote. As with every aspect of his life, if he did it at all, he did it thoroughly.

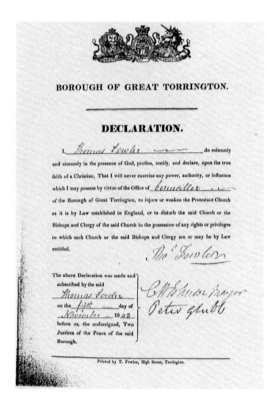

In 1835 he took on another post, Inspector of Weights and Measures, an important if not universally popular role. It was back in 960AD that Edgar the Peaceful first decreed that all measures must comply with standards kept in London and Winchester. This decree was extended in the reign of King John when a national standard of weights and measures was incorporated into the Magna Carta. Then, in 1824, the Weights and Measures Act established the Imperial system.

Mastering the complete repertoire was an awe-inspiring task. In addition to the familiar farthings, florins, pounds, ounces, gallons and pints, Fowler had to get to grips with chalders, drams, grains, lasts, pecks, poles, rods, sacks, scruples, tods, weys and more — all measures of distance, weight, length, area, volume, temperature or currency with variously obscure origins. It is said that the English yard was taken as the distance between Henry I's nose and the tip of his outstretched arm. And the basic unit of weight was the grain — but not any grain; it must be barley. With 7000 grains to the pound, market day must have been quite a trial for the medieval trader.

It was Fowler's duty to ensure traders abided by the standards. Presumably this included regular **checks** on the pints and gills being served at the twenty-two Inns within Torrington town centre alone.

A spiritual diversion of a different kind was the sundial he delineated in 1832 for the east wall of the parish church It is a curious mix of styles. As well as the inscription *Shadows we are and like shadows we depart* there is a Greek inscription *Let the empty shadow teach thee wisdom* the Hebrew word, *Jahveh* [Jehovah] and Roman numerals.

In the same year, Fowler was successful with another post, one that was to reawaken his passion for invention. A recent Act of Parliament had overhauled the old Poor Law system. Where it was well administered, it had been effective, supporting families in their own homes or taking the elderly and sick into poor houses. But by Fowler's day, need was growing faster than the country's ability to support it. Many thought the relief over-generous, encouraging idleness and irresponsibility and creating too great a burden on decent hardworking members of the community.

Poor Law reformers pushed through new legislation sweeping away the old system and replacing it with a much harsher regime.

Workhouses, at the core of the new system, were actually designed to be "uninviting places of wholesome restraint" a travesty parodied by Dickens in *Oliver Twist* where the sage members of the board:

> ... Established the rule that all poor people should have the alternative (for they would compel nobody, not they) of being starved by a gradual process in the house, or by a quick one out of it. With this view they contracted with the waterworks to lay on an unlimited supply of water; and with a corn-factory to supply periodically small quantities of oatmeal; and issued three meals of thin gruel a day, with an onion twice a week, and half a roll on Sunday.

The introduction of the punitive workhouse system was just one of many changes that had local people up in arms — literally. The poor of Sheepwash, a village just outside Torrington, were enraged when poor relief, previously given as money, was replaced by bread. They caused such a riot that the North Devon Yeomanry Cavalry were called out to disperse the mob. But whatever the rights and wrongs of the new law, the twenty-three parishes surrounding Great Torrington were united to form the new Torrington Poor Law Union. And the person responsible for calculating the cost of providing food or medical care in paupers' homes or keeping them in the Poor Law school, workhouse or asylum, was Thomas Fowler.

Mathematics had been Fowler's life for so long that he took financial calculations in his stride. But even with a slide rule and printed tables, he despaired at the monumental task every Poor Law treasurer faced. The calculations were horrendously complicated and undetected errors in the pre-calculated tables wasted valuable time.

Scientists, tradesmen, astronomers, navigators, bankers, engineers, surveyors, and others from all walks of life regularly consulted tables. Anyone in the money lending business needed tables on the interest payable at set rates. Navigators used them to establish their exact position at sea; any error here and lives and a hold full of valuable cargo could be lost as ships foundered on shallow reefs or rocky coastlines.

But errors there were. They crept in during the four stages of production, from calculation through transcription and typesetting to proofreading. Someone had to perform all the calculations by hand to begin with, an unbelievably tedious task. Two computers — at this time meaning people not machines — then cross-checked the calculations and the results were copied by hand into lists for the printer ready for typesetting, a particularly vulnerable stage. Compositors worked with endless pages of meaningless numbers — no chance to check accuracy with a quick glance. This applied equally to the proofreaders looking at each individual digit in page after page of mind-dulling similarity. The common factor throughout was human fallibility. As long as tables continued to be produced by heads and hands, errors were unavoidable. All mathematicians were aware of this but one in particular, Charles Babbage, was determined to find a solution.

Babbage was born in 1791 in London but, like Fowler, came from a Devon family and retained strong links with the county. Coincidentally, he also studied through the night from *Ward's Mathematicks,* but that is where any similarity with Fowler's early life ends. Babbage's father was a prominent London banker with both the means and the inclination to provide his son with the best possible start in life. While Fowler's life unfolded within the confines of Great Torrington, the young Charles Babbage was out and about in London, enjoying a childhood full of rich and varied experiences. One memorable visit was to the London museum of John Joseph Merlin, an inventor and automaton maker. On a tour of Merlin's workshop, Babbage was fascinated at the way his mechanical figures mimicked human actions. The man now regarded as "the father of computing" was passionate about his subject from an early age.

Education at a series of private schools culminated in a place at Trinity College, Cambridge. Babbage was fortunate in being born to parents who had both the will and the means to send him to University. Many years later, Hugh Fowler was to say of his father "There was no one, alas! to take him by the hand, and help him to carry on his studies at Cambridge, where alone such talent as he undoubtedly possessed could either have been fully developed or adequately rewarded; for that he would have distinguished himself at the University there can, I think, be no question. So he was left, without help or sympathy, to his solitary studies."

Babbage soon tired of the staid curriculum on offer and his tutors' reluctance to embrace new ideas. His response was to launch *The Analytical Society*, a new forum for debating mathematical theory. Not that he spurned all aspects of college life. Babbage was an enthusiastic member of the sixpenny whist group and frequently enjoyed trips into the Fens on his London-built boat. To make these outings complete, Babbage would order the cook to send a well-seasoned meat pie, a couple of fowls and three or four bottles of wine for a hamper, an enjoyment of the finer things in life that provides a stark contrast to Fowler's daily routine.

However, Babbage was about to demonstrate a similar determination to progress practical science. In the 1820s, his frustration at error-ridden tables boiled over into action and he began to devise a mechanical solution. He produced drawings and then a working model of an automatic machine that could reliably produce accurate results. Two years later he wrote to the prestigious *Royal Society*, an organisation dedicated to promoting excellence in science. He proposed his Difference Engine — so called as it was based on a mathematical system called The Method of Differences — as a solution to the problem of inaccurate tables. He concluded "whether I shall construct a larger Engine of this kind, and bring to perfection the others I have described, will, in a great measure, depend on the nature of the encouragement I may receive." Babbage had lobbed the ball firmly into their court. It was now down to the Royal Society to decide whether or not to recommend his invention for funding.

The government was reluctant to give grants for inventions, taking the view that if new devices were any good the public would provide the finance by buying them. However, on this occasion, the Treasury was swayed by a recommendation from the Royal Society. "Mr. Babbage has displayed great talent and ingenuity in the construction of his Machine for Computation, which the Committee think fully adequate to the attainment of the objects proposed by the inventor; and they consider Mr. Babbage as highly deserving of public encouragement in the prosecution of his arduous undertaking." Within a year, Babbage received a substantial sum from the government that, supplemented by his private funds, enabled him to begin work on his new mechanical calculating machine.

Twelve years later Thomas Fowler started out his own, very different, journey towards revolutionising the process of calculation. Two men sharing a common passion — with the fate of one already inextricably linked to that of the other.

Fowler's journey began with the challenge of simplifying the Poor Law calculations, an undertaking crying out for a fresh approach. With a currency divided into pounds, shillings, pence and farthings, the English monetary system was desperately complicated. At the beginning of each calculation Fowler needed to convert each sum into farthings and back again at the end. With 4 farthings to the penny, 12 pennies to the shilling and 20 shillings to the pound — a total of 960 farthings to the pound — this created two headaches: repeated multiplication and division and calculations involving enormous numbers.

Until now, computers, the human kind, had sighed deeply, sharpened their quills and simply got on with it. Not Thomas Fowler. It is almost possible to see him rubbing his hands in relish at the challenge of finding a simpler solution.

In common with every mathematician of his day, Fowler relied on tables using the decimal format, a method of making numbers work that had become part of the fabric of society. From its earliest origins in India, as far back as 3000BC, the decimal system had evolved into the familiar symbols 0, 1, 2, 3, 4, 5, 6, 7, 8 and 9, carrying to the tens, hundreds, thousands and so on after each unit of 10. Perhaps decimal became rooted in our society because we have ten digits on our hands but whatever the reason, it took the inventive mind of Thomas Fowler to veer away from convention and take a different view.

The result, as Fowler himself says "of constantly searching after some other method more simple and of easier application," was a staggering leap of imagination. Having shut himself away from all distractions he took the first step. "Happily, I hit on the idea that any number might be produced by a combination of the powers of two or three."

There it is again, that phrase "I hit on the idea." But this was no lucky break or random discovery. His decision to leave the decimal world behind and explore two different numbering systems, binary and ternary, was a calculated application of his mathematical genius that anticipated the same seminal leap computer pioneers revisited almost a century later. Professor Ralph Merkle, Senior Research Fellow at the Institute of Molecular Manufacturing in California, recently shared his view that "Computers might have changed history and our world almost a century sooner had the ideas of Fowler been understood and adopted by Babbage." The technology may have taken longer to catch up but imagine the implications of the underlying principles being embedded in society in the 1800s.

While binary and ternary may be a foreign land for many readers, Fowler would have been comfortable with both. Briefly, the binary system is based on two and uses only the digits 0 and 1. Ternary is based on three and uses only 0, 1, 2. (Instead of decimal using 0, 1, 2, 3, 4, 5, 6, 7, 8 and 9) Considering either was a seismic departure from convention in pursuit of simpler calculation.

He experimented with both binary and ternary, producing lengthy tables in both that resulted in a massive saving in time and effort in every calculation he made.

Fowler was characteristically self-effacing about his discovery, acknowledging that mathematicians had known about the theory for some time. Maybe, but no one had made the leap of imagination to apply it as he had. His appreciation of his discovery was almost poetic. After months of wrestling with this supremely difficult problem he declared the solution "an obvious and simple and a truly beautiful mode of keeping the Accounts of the Poor Law Unions." His enthusiasm for this potentially dry subject shines through, undiminished since he first discovered the magic of mathematics as a young man.

Recognising the impact his Tables could have, Fowler brought them to the attention of the Board of Guardians. The Chairman, Lord Clinton, and Charles Johnson, the auditor of the Union accounts, immediately pressed him to publish. By the end of 1838, *Fowler's Tables for Facilitating Arithmetical Calculations intended for calculating the proportionate charges on the Parishes in Poor law Unions and which are also useful for various other purposes*[8] were available. Fowler's Tables transformed his day-to-

---

[8] See Appendix 2

day work on the Poor Law accounts, but he was not about to rest on his laurels. Convinced the whole process could still be improved he set out on what he later described as "untrodden ground." He began to explore the possibility of adapting his binary and ternary arithmetic for calculations made not by hand, but by machinery. Thomas Fowler had the construction of a unique mechanical calculating machine in his sights.

TABLES

FOR FACILITATING

ARITHMETICAL CALCULATIONS,

INTENDED

FOR CALCULATING THE PROPORTIONATE CHARGES
ON THE PARISHES

IN POOR LAW UNIONS,

AND WHICH ARE ALSO

USEFUL FOR VARIOUS OTHER PURPOSES.

DEDICATED BY PERMISSION TO
THE
RIGHT HONOURABLE LORD CLINTON,

CHAIRMAN OF THE BOARD OF GUARDIANS OF THE
TORRINGTON UNION.

BY THOMAS FOWLER,
TREASURER TO THE TORRINGTON UNION.

LONDON:
PRINTED FOR LONGMAN, ORME, BROWN, GREEN, AND LONGMANS,
PATERNOSTER-ROW.
1838.

Chapter 4
[+ +]

A Brief Diversion

The very human stories behind the early calculating devices and their place in the sequence of events that led to Fowler's invention are strangely compelling. But if you prefer, you can fast forward to *Thomas Fowler's desktop device is conceived*, towards the end of this chapter, to continue his story.

The Contortions of Chinese fingers

Humans have always relied on aids to calculation. Before mechanical devices there were fingers. The early Chinese used the three joints on each finger. The thumb nail on one hand would touch each joint on the other, first on one side of the finger, then along the top, then down the other side, giving nine different marks. With each finger representing respectively units, tens, hundreds and thousands and the thumb, hundred thousands, with an element of contortion, a million is achievable.

The Victorious Abacus

One of the earliest and most enduring of solutions is the Abacus, a mechanical aid that evolved over more than two and a half thousand years. In its earliest incarnation the abacus was a simple counting board, probably no more than lines drawn in the sand with pebbles used to represent the numbers; ideal for on the spot transactions but not exactly durable. During Greek and Roman times the boards became more sophisticated, evolving into metal or stone slabs with carved grooves and bone, ivory or even silver coins replacing the pebbles. They gave each marker the name calculus, using the verb calculare to refer to operations performed with these markers — hence the English verb, to calculate.

The modern Abacus appeared around 1200AD in China and was later developed by the Japanese. The carved grooves and loose pebbles disappeared, to be replaced by the familiar rectangular wooden frame holding a series of rods on which freely moving beads are clicked back and forth. Although an aid to memory rather than a calculating device as such, it captured the market.

It is tempting to relegate the abacus to history but in 1996 it appeared in its most astonishing reincarnation. Research scientists at IBM constructed the world's smallest abacus with individual molecules representing the beads. To count from 0 to 10, rows of ten molecules are pushed back and forth by a conical needle terminating in a single atom in a similar fashion to beads on an abacus.[9]

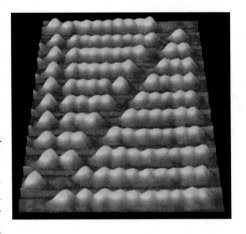

The traditional abacus was also holding its own well into the twentieth century. On November 12th 1946, Mr. Kiyoshi Matsuzaki, champion operator of the abacus from the Ministry of Postal Administration in Japan, issued a challenge to Private Thomas Nathan Wood, an expert on the electro-mechanical calculator, serving with General MacArthur. Amazingly, in addition and subtraction the traditional abacus emerged the undisputed victor. The US Army

[9] Courtesy of James K.Gimzewski FRS, UCLA

newspaper, *Stars and Stripes*, reported "the machine age took a step backward yesterday at the Emie Pyle Theatre as the abacus, centuries old, dealt defeat to the most up-to-date electric machine now being used by the United States Government." The abacus victory was decisive. But not on all counts. With division and multiplication, the abacus began to show its limitations as a mnemonic device designed to keep track of what is largely a mental calculation.

Napier's Bones — a compact with the devil?
The challenge to devise a mechanical aid for multiplication preoccupied inventors for centuries. In 1614 Scotsman John Napier said of his invention of Logarithms, a new type of mathematical table to aid calculation "Seeing there is nothing (right well-beloved Students of the Mathematics) that is so troublesome to mathematical practice, nor that doth more molest and hinder calculators, than the multiplications, divisions, square and cubical extractions of great numbers, which besides the tedious expense of time are for the most part subject to many slippery errors, I began therefore to consider in my mind by what certain and ready art I might remove these hindrances".

Two centuries later, Napier's Logarithms were an invaluable aid for Fowler. But it was another invention, Napier's Bones, that was the mechanical aid. The name conjures up dismembered skeletons, a thought some were happy to embrace with rumours circulating explaining Napier's extraordinary abilities as evidence of a compact with the devil and his study of the black arts. The truth is less sensational. The name simply

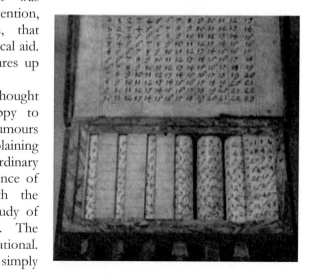

derives from Napier's choice of ivory, giving the appearance of bones. They were actually rods marked with numbers so they could be used for multiplying, dividing and taking square roots and cube roots.

But, like the abacus, Napier's bones were more of a mechanical tool than a device, a response to the discovery that addition can be short-circuited with multiplication tables. When we multiply 8 by 5 we add 8 five times. But we can then dispense with the process of addition by learning the result, 40, in a multiplication table. Napier's bones save us the task of memorising these tables but still involve the operator in several additions.

Da Vinci's Device — the first mechanical calculating machine?
The first attempt to design a calculating machine appears to have come from a familiar figure, Leonardo da Vinci. As well as being an eminent painter, he was a gifted sculptor, architect, engineer, inventor and all round scholar. He was at one moment the geologist, then anatomist — personally dissecting bodies, a gruesome experience in these pre-refrigeration and formaldehyde days — botanist, embryologist or weapons designer, a strange addition for someone who abhorred war. But the missiles, machine guns, grenades, portable bridges, massive catapults, tanks and submarines he designed helped to keep him solvent.

His output included designs for a telescope, floating snowshoes enabling him to walk on water and a mechanical lion. More famously he unravelled the science of flight with the aid of an artificial bird. Significantly, da Vinci brought a new approach to his scientific investigations. He was meticulous in his observation and systematic in recording his findings, establishing a method of scientific enquiry that has endured for centuries. At the core was his insistence on recording things by printing, a process he engaged with a full three hundred years before Thomas Fowler. And from a comparatively recent discovery, we know da Vinci shared another of Fowler's passions. When two of da Vinci's manuscripts, known as the Codex Madrid, were rediscovered in 1967 they were found to contain a drawing of a calculating device.

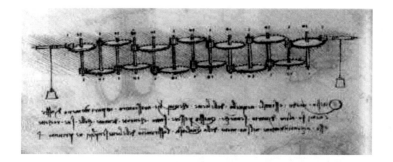

Dr Roberto Guatelli, an expert on da Vinci, recognised it as similar to a drawing in another manuscript, the Codex Atlanticus. From the two he built a replica of Leonardo's mechanism consisting of thirteen wheels, each registering a unit from 0 to 9 and moving up through units, hundreds, thousands, etc.

However, Dr Guatelli's imagination may have exceeded that of da Vinci. It has been suggested the reconstruction is not accurate and the machine de Vinci sketched could not be built. Attempting to replicate the ideas of someone long dead is a treacherous process; assessing the success or otherwise of those who have attempted it can be equally troublesome.

The Tragedy of Schickard's Machine
Another contender for inventing the first mechanical calculator is Wilhelm Schickard, a German mathematician. Coincidentally, it also took three centuries for his claim to be established. Schickard's

machine, invented around 1642, was unknown until his detailed notes were rediscovered and used to construct the machine in 1960.

On September 20th, 1623, Schickard wrote to his friend Johannes Kepler, a noted astronomer. "What you have done by calculation I have just tried to do by way of mechanics. I have conceived a machine consisting of 11 complete and 6 incomplete sprocket wheels. It calculates instantaneously and automatically from given numbers, as it adds, subtracts, multiplies and divides."

Essentially he combined the concept of Napier's Bones with a simple adder. Tragically, on February 25th, 1624 Schickard then wrote "I had placed an order with a local man, Johan Pfister, for the construction of a machine for you; but when half finished, this machine, together with some other things of mine, specially several metal plates, fell victim to a fire which broke unseen during the night, three days ago. I take the loss very hard, especially, since there is no time to produce a replacement soon."

The concept Schickard employed was inspirational but his machine was never rebuilt and its potential impact on the history of calculation was lost. A fact brought home by the design of the next significant device to take centre stage.

The Pascaline — ahead of time and technology

Before the da Vinci and Schickard machines came to light, Frenchman Blaise Pascal was credited with inventing the first mechanical calculating machine. Pascal's interest in mathematics appears to have been a case of forbidden fruit. His father, Etienne Pascal, banned his son from studying mathematics until he turned fifteen, only relenting when Blaise demonstrated a remarkable ability for geometry by the age of twelve, a precocious talent he later applied to the invention of his calculating machine.

His machine, the Pascaline, was a significant step forward in the creation of a device to meet a practical need. As with Fowler, it all started with a new job. In 1639 Etienne Pascal was appointed as a tax collector and found himself faced with tedious, lengthy calculations. His son's response was a labour saving device designed to do the work of six men.

The mechanism was built on a brass rectangular box where a set of wheels, initially 5 but later 6 or 8, were rotated in a clockwise direction with the use of a pin. Significantly this machine also included automatic carries.

After obtaining the equivalent of a patent for his device in 1649 he experimented with different methods, generating fifty prototypes over ten years. But the available technology was not sophisticated enough to enable Pascal to craft his mechanism with the precision it required. The machines acquired a reputation for unreliability and few were ever sold.

The Stepped Reckoner — freeing slaves to the labour of calculation
The next significant innovation was unveiled by German mathematician, Gottfried Wilhelm Von Leibniz in 1672. Leibniz grew up in Leipzig, and progressed through school and university at

frightening speed. By the age of twelve he had mastered Latin and Greek and was avidly studying metaphysics and theology. At seventeen he graduated from university, having studied philosophy, mathematics, rhetoric, Latin, Greek and Hebrew, and went on to study for a doctorate in law.

This was only the tip of the iceberg. Leibniz had an astounding capacity to turn his attention to anything. It was even one of his aims to collate all human knowledge, something he distilled down to the more manageable ambition of bringing together learned societies to co-ordinate research into their various areas of interest.

One enduring topic was his interest in mathematics. Specifically he dreamt of developing a symbolic language so that "the truth of any proposition in any field of human inquiry could be determined by simple calculation." This ambition propelled him into an exploration of mathematical notation and the invention of the calculus — bringing him into direct conflict with Newton. Newton believed Leibniz had stolen his methods, an accusation Leibniz hotly disputed.

In the early 1670s, Leibniz began work on his calculating machine. He believed that "it is unworthy of excellent men to lose hours like slaves in the labor of calculation which could safely be relegated to anyone else if machines were used." A sentiment Fowler would have thoroughly approved of. Leibniz even called his machines living bank clerks. His Stepped Reckoner was designed to deal with numbers up to twelve digits and handle addition, subtraction, multiplication and division. The unique component distinguishing it from Schickard and Pascal's machines was his stepped drum, a novel incrementing gear mechanism.

Leibniz worked on two prototypes, concentrating on speed and accuracy. But he could not overcome the same problem that defeated Pascal; the failure of the available technology to create the precision these visionary devices required. Tragically, Pascal and Leibniz's machines shared a common fate — destined to remain curiosities rather than practical, everyday calculating devices.

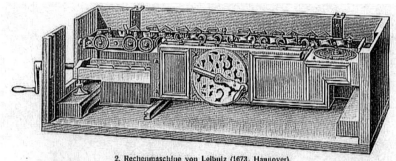

2. Rechenmaschine von Leibniz (1673, Hannover).

3. Leibnizsche Rechenmaschine, geometrische Zeichnung.

Babbage's Difference Engine takes shape

Over the next hundred and fifty years others entered the fray, building on these early designs. Then, in the 1820s, two significant devices emerged — Babbage's Difference Engine, with lofty mathematical ambitions, and the Arithmometer, with more modest aspirations. The Difference Engine was a concept originally explored by German engineer, J. H. Muller, in 1786 but then forgotten. Babbage's machine, that began to take shape in 1822, was designed to compute lengthy tables, setting the answers directly in type, effectively eliminating the errors introduced in the calculation and printing processes. More of Babbage's machine later.

The Arithmometer makes it to market.

The Arithmometer, invented by Charles Xavier Thomas de Colmar in 1820, was a practical aid to calculation, capable of performing the four arithmetical operations in a simple and reliable way. Uniquely, it achieved exactly what it set out to do. The prototype demonstrated the usual problems with reliability but over the next thirty years, de Colmar benefitted from improvements in technology until he arrived at a simple, easy to use, accurate version of his original design.

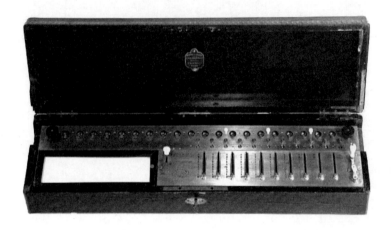

By 1845 Thomas de Colmar had established a business based on sales of the Arithmometer, the first calculating device to make it as a commercial success. Sales publicity described "an instrument by means of which persons the least familiar with figures are enabled to work out all the rules of Arithmetic, and men of Science are enabled to solve the most intricate problems in a few seconds; it is simple and strong, and easy to manipulate … eight figures can be multiplied by eight figures in eighteen seconds; sixteen figures be divided by eight figures in twenty seconds; and a square root of sixteen figures be extracted, with the proof, in less than two minutes. With this instrument great saving of time is effected, half an hour only being required to obtain the same result as by a long day's work with the pen, and that without fatigue and with mechanical accuracy."

Despite being an expensive piece of office equipment, around 1500 Arithmometers were sold to insurance offices, observatories, railway companies, and sundry government departments, proving beyond doubt that a market existed for a desktop calculator.

Thomas Fowler's desktop device is conceived

In 1838 the Arithmometer had yet to make its mark and although Babbage had started work on his Difference Engine, this was never intended to be a desktop machine. With over 25000 parts, predicted dimensions of eight feet by seven by three and weighing several tons it was anything but. The market was still wide open for a convenient, reliable, affordable desktop calculator to transform the working day. And the device taking shape in Fowler's mind, a device that operated on a completely different principle to everything that had gone before, was a prime candidate.

As usual, once the possibility presented itself there was no letting go. Fowler had set his sights on a machine that could transform the lives of clerks across the country, across the world even. Nothing could keep him from his workshop with such an achievement tantalisingly within reach.

He now had a crucial decision to make. Which of the two scales used in his Tables was right for the machine: binary or ternary? Both had advantages over decimal. Calculation using a decimal machine was tediously long-winded. To multiply 98 by 76 using an Arithmometer required fourteen actions. Imagine the labour involved with larger numbers. With simplicity his byword, Fowler was looking for a device that could multiply with far fewer actions for each digit. Known today as a direct multiplier, it has two significant advantages — it is quicker and there is far less chance of error.

Fowler soon realised either scale would work. But which to choose? Binary was the ultimate in simplicity with only two choices for each digit instead of three in ternary. But this simpler move has to be performed more often, meaning more moving parts, creating a larger machine. With a desktop device in mind this was pushing him in the wrong direction. There were also serious issues with negative numbers and fractions with binary. The Poor Law accounts were littered with fractions, so Fowler opted to find a better solution rather than live with an unsatisfactory compromise. With binary out of the picture he took the giant leap into the world of ternary.

Fowler's appointment as Inspector of Weights and Measures just three years earlier may have provided the inspiration. The pictures below show the Scales used by Fowler to check the accuracy of the weights used by traders on market day; weights that came to have a very particular significance for him.

Finding the most efficient combination of weights is a scenario mathematicians will be familiar with as Bachet's Weights Problem. 'What is the least number of pound weights that can be used on a scale pan to weigh any integral number of pounds from 1 to 40 inclusive, if the weights can be placed in either of the scale pans?' The answer? Four weights of 1, 3, 9 and 27 pounds respectively, all part of the ternary scale. With these, traders could accurately weight anything up to forty pounds. Simplicity itself.

The number 3 has a proud history. From the Holy Trinity to the three legged stool it is a remarkable digit. It is the number of time — birth, life, death; beginning, middle, end; past, present, future. The number of belief — Brahma, Vishnu, Shiva; Father, Son and Holy Ghost, Buddha, Dharma and Sangha. It is even the number of fairy tales — the three little pigs, three wishes, three billy goats gruff. We perceive three dimensions and conceptualise in three's — body, mind, spirit; thought, word deed; yes, no, maybe. Three even pervades our superstitions with the conviction that bad things come in threes!

Fowler was breaking new ground focusing on ternary but so far his achievements had relied on the groundwork put in by other mathematicians. There is no denying his brilliance in doing this, but Fowler was now intent on innovation as well as application.

At the very end of the introduction to his Tables, Fowler offers an intriguing taster. "Should the sale of the present Edition be favourable, another will soon follow, in which the Ternary Table will appear under another Form, and extend from Unity to Numbers almost indefinitely great, and also contain some other curious and I hope useful matter." What were these intriguing references to another form and "some curious and I hope useful matter"? Fowler explains. "In the course of my observations on the Binary and Ternary Scales, I have fallen on a species of Binary and Ternary Arithmetic which appears to possess some curious properties, but, as writing on this subject is incompatible with my present purpose, I have only given a short Example of Multiplication, in what may be termed Ternary Arithmetic, the process is extremely easy, and may be extended to very large numbers."

Extremely easy? For Fowler perhaps! Interpreting this example is challenging even for a seasoned mathematician; impossible for the casual reader.

There is more in Appendix 4 for anyone interested in the arithmetic, but essentially this table illustrates Fowler's discovery of a new mathematics, now called signed ternary. In an unprecedented step, he discovered he could use three symbols, minus (−), zero (0) and plus (+), instead of numbers. A brilliant innovation, incomprehensible at first glance but once interpreted, a giant leap towards quick and easy calculation, removing the need for any mental arithmetic whatsoever.

The significance of this development cannot be overstated. In stark contrast to existing complicated mechanisms, Fowler had discovered the perfect compromise between simplicity of operation and economy of parts in a mechanical device. He was on his way to creating a machine unique in its theoretical basis, design and construction — pedestrian words concealing a breathtaking development.

Chapter 5

[+ − −]

Lateral Links

Deciding between binary and ternary was not Fowler's only dilemma. Fourteen years after his bitter experience with the Thermosiphon, Fowler was in a quandary. Employing someone to help build his new invention would speed up construction, but it also meant sharing his inspiration. His almost visceral reaction to this leaps off the page. One betrayal was enough. There was no way he was going to risk anyone pirating an invention that he believed would be his greatest claim to fame.

Fortunately, Fowler's choice of ternary meant that the parts and precision needed in Babbage's machine were unnecessary. He could construct his machine out of wood. The years he had spent working alongside his father in the cooper's yard now came into their own. With his woodworking skills at the ready and materials available locally, he set to work.

The picture in Babbage's workshop was very different. He positively embraced delegation. "Having myself worked with a variety of tools, and studied the art of constructing each of them, I at length laid it down as a principle that, except in rare cases, I would never do anything myself if I could afford to hire another person who could do it for me." Freed from the labour involved in the day-to-day construction of his machine, Babbage completed hundreds of pages of notes and drawings. He also had the time and the means to travel, visiting manufacturers at home and abroad to extend his knowledge on

45

machine processes that could be adapted or reinvented to meet the new technical requirements of his Difference Engine.

But this level of personal investment was unsustainable. Babbage's machine required hundreds of near-identical parts. Before the age of mass production these had to be made by hand, a costly process. The man in charge, Joseph Clements, was highly skilled but Babbage himself conceded that Clements' charges were "inordinately extravagant".

Babbage's attempts to claim reimbursement from the government plagued him for the rest of his life. Support from Francis Baily and Sir John Herschel, both eminent astronomers, resulted in occasional payments, but Babbage was up against a government who believed they were assisting "an able and ingenious man of science, whose zeal had induced him to exceed the limits of prudence."

Work on the Difference Engine was suspended, recommenced, suspended; then in 1830, with difficulties over finance apparently finally resolved, resumed. Two years later Joseph Clements completed a fully functioning portion of the Difference Engine, the world's first automatic calculating device.

This was an astounding achievement. All other devices required the continuous, informed intervention of an operator. Now anyone could crank a handle and read off the results; there was no need to understand the workings or the mathematical principles underlying them. This working portion contained over 2,000 individual parts and was a magnificent example of precision engineering. Without Babbage's single-minded determination to continue this work, the support of his friends and colleagues from the Royal Society, and a Government grant to date of £9,000, (around £720,000 today) even this section of his innovative device would not have been built during his lifetime.

Thomas Fowler was attempting to realise his vision with locally obtained timber, the facilities of his small workshop, and his own hands. And his was not a well-worn path. He had already veered from the traditional in choosing ternary over decimal but this was only the beginning. Instead of echoing the familiar design of earlier devices, and Babbage's Difference Engine in particular, Fowler now made another leap of imagination with the fundamental construction of his machine.

First impressions of the device taking shape in Fowler's workshop could not have been further from Babbage's gleaming, synchronised mass of rotating metal wheels — a triumph of engineering, admired and appreciated by the most uninformed spectator as an artefact if not for its wider significance.

Fowler's machine was made of wood: no shine, no glamour.

Next in line to take the casual observer by surprise was the shape. The Difference Engine had a four-square solidity to it, with a dense core that reinforced initial impressions of substance and significance.

By contrast, Fowler's machine had an almost skeletal appearance. In place of a dense mass of interlocking cogs and wheels, Fowler's system was based on the principle of sliding rods. This change of tack was yet another example of his flair for finding simple solutions to complex problems. One of Fowler's local sponsors Sir Trevor Wheler was later to remark "nothing can be simpler than its [the machine's] construction."

Where did his inspiration come from? There are clues in one of the first reviews. "The machine itself is on the principle of the old abacus, or calculating rods." In casting around for alternatives to the rotary model used by Babbage, de Colmar, Schickard, Pascal and Leibniz, Fowler may well have contemplated the centuries old abacus where markers moved in a linear fashion, acquiring different values depending on their position. In Fowler's construction, the rods themselves were the markers, acquiring different values depending on where they rested; a different construction but fundamentally the same principle.

The reviewer's opinion that Fowler's machine "has somewhat the appearance of a pianoforte, or organ, with all its keys laid bare," is particularly interesting. Not only was Fowler a long standing organist at the parish church, reinforcing the common link between maths and music, but some years earlier he had added organ building to his many other talents.

In the eighteenth century, individual wind instruments and voices provided music for church services. But in 1809, Lord Rolle donated a barrel organ to Great Torrington church. The congregation could now enjoy music whether or not musicians were present. But it was not a popular option. When he was appointed organist, Fowler set to work adapting the rods and linkages to function manually. So not only was he able to play the instrument, he was also intimately acquainted with its construction and operation. And he had had a recent reminder of the organ's workings. As he worked late in his workshop one night, the wind outside gusted at almost hurricane force. Deep in thought, Fowler would have been aware of the storm but unconcerned — until the sound of falling debris became too close for comfort. No more than a few yards from his house, the top of the church steeple became dislodged, falling onto the roof below.

The devastation was clear the following morning. Masonry had fallen onto the west gallery where the organ was housed. When Fowler worked his way into the bowels of the mechanism to check for damage, perhaps his intimate knowledge of its construction suddenly took on a new light. He was surrounded by rods, levers and linkages, all with the specific purpose of translating the action of one part of the mechanism to that of another. Was this another time when Fowler 'hit on an idea'?

There is another direct route to the idea of sliding rods from Fowler's development of ternary arithmetic. It is possible to get a feel for how these functioned in the machine without delving into the arithmetic (contained in appendix 4) by imagining the lighter grey bars in this illustration as sliding rods. The position of the bars indicates whether a number is to be added, subtracted or left out of the calculation completely. In this example, 27 is to be subtracted, 9 and 3 added and 1 ignored, giving −15. For a different calculation the bars, or rods, would simply be slid along the rows, or grooves, to different positions.

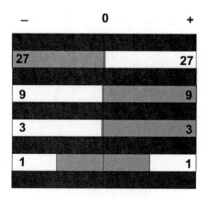

Whatever the source of his inspiration, Fowler now pressed on with bridging the gap between concept and reality. The framework was relatively quick and easy to construct, but replicating the moving rods and levers was more time-consuming. Fowler worked on throughout the winter until the moment came when, in the spring of 1840, he placed the last rod alongside the others that stretched across the width of the frame before him.

Eighteen months after he first 'hit on the idea' of a unique calculating machine, after all the long, solitary hours in his workshop, there it was, finally complete. A moment of quiet satisfaction? More probably the cue for an invasion by the entire family. Sons Hugh, Henry and Charles were no longer at home but his wife, Mary, and remaining children, Caroline, Cecilia, Frances, Newell, Silas and Paul, would have been treated to a demonstration by a proud husband and father. Congratulations may have been tinged with relief. Finally, the work that had kept Fowler shut away in his workshop was done. At last the family might get some help with the printing office and the machine itself might mean fewer hours taken up by Poor Law business.

Fowler's invention certainly met all his expectations when it came to the Poor Law calculations. To his absolute joy, once the various parts were set and the levers moved according to a few predetermined rules the calculation was complete. The machine was still too large, six feet by three by one, but even in this form it was much smaller than Babbage's completed Difference Engine would have been.

Reconstruction of Babbage's Difference Engine – Science Museum 1991

However, a direct comparison between the two machines is difficult as the Difference Engine had a printing mechanism attached — tables were first calculated then printed. Clerks were never intended to use the device, just the tables it produced. The ability to print tables was an unnecessary complication for Fowler's calculator, designed to sit on a desktop and solve calculations for individual office workers. But even if he had attached a printer, his machine would still have been a fraction of the size, and cost, of a completed Difference Engine, particularly when constructed in metal.

It was Fowler's choice of ternary that made these benefits possible. He later wrote "had the ternary instead of the denary [decimal] notation been adopted in the infancy of Society, Machines something like the present would long ere this have been common, as the transition from mental to mechanical calculation would have been so very obvious and simple."

But for his calculator to become widely used, unskilled office worker's needed to know how to operate it. With Babbage's machine all they had to do was turn a handle. Fowler's was a very different proposition. How many clerks knew about signed ternary? The obvious solution was to produce tables converting decimal into ternary. Then all an operator had to do was find the number in decimal, trace across for its ternary form, a variation of +, 0 and −, and enter this code into the machine. They then operated the machine according to a few predetermined rules and converted the result back into the decimal by using the table in reverse. All they needed to know was the procedure for operating the machine, not the mathematical principles behind it. Clerks could now perform any calculation without knowing the first thing about arithmetic, ternary or otherwise. With a small manual describing the procedure, the conversion tables and some commercial backing, Fowler's machine was ready to take its place in offices across the country.

Surprisingly for such an entrepreneur, this was not the route Fowler took. He was thrown back to earlier times when, instead of a life at university debating mathematical theory in hallowed halls, he had studied alone after long days in the fellmonger's yard. Yes, he had become a successful businessman, banker and councillor, but this success was no more than an impressive frame for a canvas that was yet to be unveiled. His real passion, the talent that gave his life perspective, was his mathematical ability. And after all these years he finally had an invention that would prove that ability beyond doubt. Or did he? This nagging uncertainty diverted him from promoting his machine in a practical way. He had to know: despite his unconventional path, could he command the respect of his country's foremost mathematicians and scientists?

The obvious place to start was with his local sponsors. Fowler's achievements had attracted interest amongst Torrington's gentry and four were about to become crucial to his future success. One morning in April 1840, Lord Clinton, a prominent landowner and himself a mathematics graduate; John Moore Stephens, initially a Torrington man but now Archdeacon at Exeter; Charles Johnson, several times mayor of the town and Sir Trevor Wheler, a local philanthropist, all eagerly attended a demonstration of Fowler's latest invention. This audience at least were quick to appreciate its significance. In fact, they were so enthusiastic that plans were soon in hand to make sure it received the attention it deserved.

There was just one problem: despite his determination to get his work recognised, Fowler still refused to release any plans or the machine itself. Twelve years on from the debacle over his Thermosiphon, he was adamant that this time he would release nothing that would enable others to pirate his invention. His stubbornness created a stalemate.

Fortunately, Archdeacon Stephens came to the rescue. He was so impressed with Fowler's machine that he arranged for him to demonstrate it to Charles Babbage — a name synonymous with mechanical calculation, and someone at the very heart of London's academic community. Despite their shared Devon heritage and passion for calculating devices, they had never met.

Charles Babbage

Finally, after all the heartache, the labour, the long, solitary hours in his workshop, Fowler was to have the opportunity to debate his machine with someone eminently qualified to assess its underlying principles and potential. Letters flew back and forth between Devon and London and a date was set for Fowler to demonstrate his machine.

As May Day dawned in Torrington, Thomas Fowler turned his back on a peaceful retirement and prepared to keep the most important appointment of his life at 47 Hunter Street, Brunswick Square, London.

This journey to London was a massive undertaking for Fowler. Born and bred in a small Devon town more than two hundred miles from London, he was an innocent when it came to city life. Apart from his journey there to lodge his Patent, it is unlikely he regularly travelled further than the small neighbouring towns of Bideford, Barnstaple or perhaps Tiverton to check on his sons' progress at Blundell's School. He was also just days away from his sixty-third birthday. The prospect of a two-day journey in a poorly sprung coach pulled by a team of horses along badly rutted roads cannot have been welcome.

If he had been travelling two years later Fowler could have taken the train from Bridgwater. It had already reached Reading and in a few months time would be open as far as Bristol. But he may still have opted for the coach. The early railways were bone-jarringly uncomfortable and had an appalling safety record, highlighted by, of all people, Charles Babbage.

Babbage's determination to create a fully automatic calculating machine was not the only challenge he set himself. He had ambitions to become a politician, wrote extensively on social reform, travelled widely, and was a prolific inventor. He invented an ophthalmoscope, the first camper van and devised the cowcatcher for clearing railway lines. Babbage also followed in Leonardo da Vinci's footsteps, literally, by experimenting with shoes for walking on water. Other ideas included designs for a diving bell, altimeter and seismograph, a submarine, the introduction of flat rate postage, and reflections on the use of tidal power. Prolific output signalling a remarkable mind.

Babbage was caught up in the excitement of the railways, but dismayed at their discomfort and potential for disaster. On one occasion he was scheduled to take an engine out to conduct some experiments, but during the night, a goods train hauling twenty-five ballast wagons ran straight into the back of it. Apparently both the driver and the stoker were fast asleep. Babbage changed his plans to a Sunday, having decided it was the only safe day to travel. Fortunately the fireman overslept, delaying Babbage just long enough to see Isambard Kingdom Brunel arrive along the very same track on which he had been about to depart. Brunel had been travelling at fifty miles an hour. A head-on collision would have been a very messy business.

A more immediate concern for Fowler was whether his machine would survive the journey. When the time came to leave, passers-by may well have been treated to the sight of their usually upright and sober bank manager anxiously checking and re-checking his precious cargo.

Travelling in stages, Fowler would have made Taunton for lunch, the ancient city of Wells for tea, and Bath for supper. After each stage, the coach would pull into the cobbled yard of an Inn to change horses; a hive of activity as grooms and servants bustled about, eager to feed and water both travellers and horses before hurrying them on their way. Occasionally there would be time for a yarn before the fire as food was prepared, particularly at the Pelican Inn in Newbury where the tale of the Newbury Coat was a favourite. In 1811 Sir John Throckmorton wagered that before sunset a coat could be made from wool growing on the back of a sheep that morning. The sheep were shorn, the wool spun, the cloth dyed and woven and the coat completed by twenty past six that evening. It's a sobering thought that while this coat was displayed at the Great Exhibition of 1851, Babbage's Difference Engine was not.

But this welcome respite would have been all too brief: the coach was usually on the road again before dawn. Even young men found it taxing. For an older man whose health was beginning to fail it must have been exhausting.

Late in the afternoon the coach and four finally approached London, skirting Hyde Park as it made for Piccadilly Circus and journey's end. Despite being travel-weary, Fowler would have lost no time in making for Hunter Street to unpack and assemble the machine. Having satisfied himself that all was well, he was free to reflect on his imminent encounter with Charles Babbage. Fowler would have known of Babbage's work from articles in the Mechanics Magazine and the

Edinburgh Review, one advantage of being a bookseller and stationer, but he had never met the man. There was no one better qualified to comment on his work, but how would Babbage respond to a potential competitor?

While Fowler travelled full of optimism, Charles Babbage had reached yet another dead-end. When he moved work on the Difference Engine to a new fireproofed workshop, his workman, Clements, demanded an increase in his already exorbitant fees. It was the final straw. Babbage had not been reimbursed by the government for over a year and had no choice but to stop all payments. Clements promptly dismissed the workmen and left. Thus began eight years of frustration. From the government's perspective, the Difference Engine was an expensive piece of machinery of dubious benefit to the nation and Babbage's constant hounding for a decision was an annoying irritation at a time of massive social problems. Spending enormous sums on something of limited value was not a priority for ministers seeking the instant solutions that would keep them in office.

Friends pressed for a response on Babbage's behalf, but governments came and went with no progress being made. In 1836 the Chancellor of the Exchequer did suggest the government approach the Royal Society for an opinion but it is just as well he was ignored. Showing a distinct lack of gratitude for the support they had offered, Babbage singled out the Royal Society for scathing criticism. In his paper, *Reflections on the Decline of Science in England, and on some of its causes*, he labelled them as amateurs unfit to advance the interests of science. It was a serious misjudgment. His refusal to nurture the right political alliances meant he was passed over as a Professor of Mathematics on two occasions, failed to win a seat on the Board of Longitude, and was turned down for the Mastership of the Mint three times; all posts he was eminently qualified to fill.

He continued to blame the government for failing to resolve the matter of the Difference Engine, but he had complicated discussions. When Clements made his dramatic departure, Babbage was temporarily deprived of his plans for the Difference Engine. During this hiatus he conceived the idea for a new and more advanced calculating machine, the Analytical Engine. From 1834 onwards he became obsessed with perfecting this more powerful, though also more complex, machine that effectively superseded the Difference Engine. Significantly, it embodied many of the principles employed in the modern digital computer. While the Difference Engine was capable only of executing the specific functions allowed by its construction, the Analytical Engine could be programmed by the user.

Babbage presented the government with an alternative: did they want him to complete the Difference Engine or work on the new Analytical Engine? This effectively seized up all communication, and he waited in vain for a decision. Instead, a letter arrived on a completely different matter — a request from John Moore Stephens, Archdeacon of Exeter, urging Babbage to attend a demonstration of a unique wooden calculating machine invented by a printer from Great Torrington in Devonshire. Thus, at the beginning of May 1840, while still waiting to know the fate of the machine that lay uncompleted and temporarily abandoned in his workshop, Charles Babbage made his way to 47 Hunter Street, Brunswick Square for his first meeting with Thomas Fowler.

Chapter 6

[+ − 0]

Poised for Recognition

> *A printer of Torrington has made and brought to London a Calculating Machine which Babbage has carefully examined and tells me he considers to posses very great merit being on a different principle to his own*

Dr William Buckland

When the Royal Society met in 1822 to consider Babbage's Difference Engine they were assessing the work of a friend and colleague. Many of them shared university days with him, or knew him by reputation if not by acquaintance. Fowler, on the other hand, had to impress a gathering of strangers, a remote and exclusive elite as foreign to him as France had been to the Devonians who fought there. He was never fazed by the theoretical or technical challenges he faced in the familiar environment of his workshop, but the prospect of presenting his machine in this company was a very different matter. His experience of the privileged classes revolved around binding their books or being the subject of their patronage. Now he intended to claim, at the very least, the respect of these learned gentlemen, if not equality with them.

The sights and sounds of London alone would have been overwhelming after the gentle pace of life in Great Torrington. The *St James's Magazine* highlighted the problem. "Russell Square is, under ordinary circumstances, a very nice place to walk in. If those troublesome railway vans and goods wagons would not come lumbering and clattering, by way of Southampton Row, through the square and up Guilford Street, on their way to King's Cross."

But it was not only the volume of carriages and wagons that would have been new to Fowler. London was chaotic and disorderly with the clamour of beggars, street traders and musicians; a misery grudgingly endured by Babbage. "During the last ten years, the amount of street music has so greatly increased that it has now become a positive nuisance to a very considerable portion of the inhabitants of London. It robs the industrious man of his time; it annoys the musical man by its intolerable badness; it irritates the invalid; deprives the patient, who at great inconvenience has visited London for the best medical advice, of that repose which, under such circumstances, is essential for his recovery, and it destroys the time and the energies of all the intellectual classes of society by its continual interruptions of their pursuits."

Amongst the instruments of torture were organ-grinders, brass bands, hurdy-gurdies, flageolets, drums, accordions, trumpets, bagpipes and the human voice in various forms. Babbage even quantified the damage. "I have arrived at the conclusion that I speak within limit when I state that one-fourth part of my working power has been destroyed by the nuisance against which I have protested."

Babbage's public protests only served to inflame things and he suffered a predictable reaction — a brass band playing outside his window for five hours! But he was unrepentant. "I was once asked by an astute and sarcastic magistrate whether I seriously believed that a man's brain would be injured by listening to an organ; my reply was, 'Certainly not', for the obvious reason that no man having brain ever listened to street musicians."

Other notable figures of London academic society on their way to 47 Hunter Street included Francis Beaufort, famous for devising the Beaufort Scale for wind speed — an accurate but completely inadequate description of this remarkable man. On active service in the navy he was riddled with sixteen musket shots and three sword wounds in one action alone. He later returned to sea, commanding an armed store ship on a survey off Buenos Aires. It was here that, tasked with making the ship's log book more concise, he developed his unique scale of wind force. His career at sea ended when wounds inflicted by pirates in another engagement off the Turkish coast forced a return to a desk job at the Admiralty in London.

Archdeacon Stephens also contacted the one person at the heart of the nation's scientific community — Spencer Joshua Alwyne Compton, the Marquess of Northampton. Northampton was a pivotal figure for two reasons. He was President of the Royal Society a body that, despite Babbage's attack, remained the pre-eminent institution for the promotion of science. But more significantly, he was well known for his social gatherings where prominent individuals met for some serious nineteenth century networking. Archdeacon Stephens had opened some very significant doors for Fowler.

Northampton immediately wrote to Sir John Lubbock, a vice-president of the Royal Society. "I have received a letter from a friend of mine, the Archdeacon of Exeter, begging me to go and see the invention of a protégé of his, a calculating machine, which he praises very highly … The Archdeacon tells me that Mr. Babbage, Captain Beaufort and some others are to see it on Wednesday next — I think you would like to see it. I think it very likely that Christie, Roget and Sabine might also like to see it." Of these last three, Peter Mark Roget would have been particularly keen to attend.

Roget is popularly known for his Thesaurus, compiled after his retirement and first published in 1852. Less well known is that he spent his remaining years developing a calculating machine. As a young man Roget also taught himself mathematics, although any similarity with Fowler's life ends there. Roget studied at a private school, went on to university, and pursued a career as an eminent physician. However, his early interest in mathematics never left him. In 1814 he presented a paper to the Royal Society on a new sliding rule he had developed. This paper led to his election as a Fellow, and in later years, as secretary, where he edited the Society Proceedings and prepared abstracts for publication. He still held this influential role as he made his way to observe Fowler's machine.

Also present was Augustus De Morgan, first Professor of Mathematics at University College, London. By 1840 De Morgan had already written hundreds of books and articles on mathematics, logic and philosophy. The prospect of impressing this particular gentleman must have been daunting for Fowler, although one family connection may have helped break the ice. De Morgan's younger brother was born at Clovelly on the North Devon coast, just a few miles from Torrington.

Sadly, what happened once these distinguished men were all gathered behind the doors of 47 Hunter Street on the morning of Wednesday 13th May, 1840 remains speculation. Despite the significance of this meeting for Fowler, no record survives. However, we do know that his explanation of the workings of the machine and the underlying arithmetic impressed this informed audience.

For readers interested in the detail, his demonstration of multiplication would have proceeded as follows.

Approaching the piano-shaped casing containing his machine at the open 'keyboard' end, Fowler is in a position to point out the four separate frames that make up the mechanism. The first frame, called the multiplicand, is nearest to him. Very loosely speaking, if you imagine a three-foot wide abacus frame laid flat but filled with moving rods instead of beads you will have a basic picture of the multiplicand. The first number to be multiplied is entered into this frame. Similar to this, but lying slightly higher at the far end of the machine, is the product frame, where the answer is displayed.

Between the two is the multiplier, a more difficult structure to describe. The most important section is a vertical ladder-like structure with just three rungs that pivot about the middle rung. This ladder can be made to connect with the rods of the multiplicand at the front, while levers attached to the individual rungs reach back to connect with the rods of the product frame at the back. In this way the multiplier effectively links the frames together. Detached from these and used to reduce the answer to its simplest form at the end of the calculation, is the carrying frame.[10]

Having used his tables to convert the relevant decimal numbers into his balanced ternary notation[11], Fowler enters the

---

[10] See appendix 13 for a modern interpretation of these structures
[11] The term 'balanced ternary' is synonymous with 'signed ternary'.

first value into the multiplicand by sliding the rods inside this flat frame backwards or forwards to indicate +, 0 or −. He then enters the other value into the multiplier frame. To make this possible, each rung is labelled. So the top most rung represents +, the middle rung, where the ladder pivots, represents 0, and the bottom-most rung represents −. To enter a value, say + 0 − (8 in decimal), Fowler attaches one lever to the top rung, the next to the middle rung, and the last to the bottom rung. To enter + + − (11 in decimal) he attaches two levers to the top rung and one to the bottom rung. The middle rung is left empty.

To begin a calculation, Fowler then moves each multiplicand rod to zero. Because of the way the different frames align with each other, this action causes the multiplier to move which in turn causes the rods in the product frame to move, displaying the answer. The carrying frame is then deployed, again according to a simple pre-determined rule, to reduce the result to its simplest form. All Fowler then had to do was read off the result, a variation of the +, 0 and − theme, and consult his tables to convert this back into decimal.

The problem with written descriptions is that beautifully simple actions begin to sound impossibly complicated. Fowler's audience would have found their first introduction to his machine a lot easier to grasp. Doubtless Fowler and Babbage then engaged in a vigorous debate on the relative merits of decimal versus ternary. Babbage had also considered both binary and ternary, but rejected them on the grounds of engineering efficiency. Lower number systems demand more mechanical parts to represent a given number. However, it is on record that he was impressed with Fowler's application of signed, or balanced, ternary.

After such an enthusiastic reception, Fowler must have felt that it was only a matter of time before his original work was recognised. In this respect, Fowler and Babbage were very alike. Both held a fundamental belief in their own ability, an ability that should attract its own reward. However, while Babbage had already experienced the consequences of refusing to promote, cajole and compromise to achieve his goals, Fowler was just setting out on this road.

His first obstacle lay in the preconceptions of his audience. Humouring country figures who presented London society with various oddities had become something of a fashion. With his country ways and broad Devon accent Fowler would have been a prime candidate for ridicule. But he was fortunate that his first audience consisted largely of men of integrity with a genuine interest in the subject and a willingness to encompass new ideas. Babbage was so impressed that he immediately arranged a meeting with the Reverend Dr Buckland, Professor of Geology and future Archbishop of Canterbury. He was well-known as a cheery, humorous man, full of eloquence, wit, and with a great enthusiasm for his subject. In his twenties he spent four years travelling on horseback throughout the south-west of England collecting geological sections. In later years he was renowned for his famous blue bag from which, for everyone's entertainment, he would produce his latest find — often fossil faeces from Lyme Regis. However, geology was not his only interest, and he eagerly awaited Babbage's visit, a meeting that took place at the Athenaeum Club in Pall Mall.

This was the age of Gentlemens' Clubs, formed to offer members comfortable venues in town. The Garrick Club offered literary men a rendezvous, the Carlton was purely political, but the Athenaeum had broader aspirations, catering for men known for their scientific or literary attainments, eminent artists, and gentlemen patrons of science, literature and the arts. The club was housed in a purpose-built mansion in the courtyard of Carlton House, a palace of a building designed in the Grecian style. It offered all the comforts of home to "Gentlemen away from home, or without a home, or even to those who preferred not to be at home." The Library was considered to be the best in London. For a fee, members enjoyed unlimited use of the books, maps, periodicals, newspapers and writing materials.

After his meeting with Babbage, Dr Buckland sat at a desk in this library, took a sheet of Athenaeum notepaper, and proceeded to write to Michael Faraday, then President of the Royal Institution.

There is a very ingenious native of Torrington, Devon, who has made a new Calculating Machine which Babbage has seen & thinks to possess very great merit. The maker's name is Fowler & he is with his Machine at 47 Hunter Street, Brunswick Square.

Should you pass that way in the next 6 days do call & look at the Man & his Machine & Believe me. Very truly Yours, W. Buckland.

Michael Faraday may well have sympathised with Fowler. After all, he was a blacksmith's son who was also apprenticed at an early age, coincidentally to a bookseller and bookbinder. But Faraday lived in London and by 1813 was ensconced at the Royal Institution as a chemical assistant. It was simply a matter of time before he became world-famous for his scientific discoveries, particularly his work on electro-magnetic induction.

Buckland decided to organise a second demonstration to allow Faraday and others, himself included, to witness Fowler's machine. He quickly dispatched an invitation to Northampton and then penned his final letter. This note, dated Wednesday 20th May, was addressed to George Biddell Airy, Astronomer Royal at the Greenwich Observatory.

A printer of Torrington Mr. Fowler has made and brought to London a Calculating Machine which Babbage has carefully examined and tells me he considers to possess very great merit being on a different principle from his own. The inventor is a man of peculiar modesty as well as great Invention in methods of calculating. Lord Northampton is going as the one to see man & his machine on Friday morning at Eleven. Should you like to join us come to 47, Hunter Street, Brunswick Square at that hour.

Unlike Buckland's other notes this one was ignored. Fowler's story might have had a very different ending had it not been.

At 11am on Friday the 22nd May, 1840, the Marquess of Northampton, Dr Buckland, Augustus De Morgan, Francis Baily, and many other Gentlemen Fellows of the Royal Society assembled at 47 Hunter Street to witness the second demonstration of Fowler's calculating machine — a prestigious gathering reflecting the interest Fowler's machine had generated. Doubtless some patronised this country mathematician, or treated his demonstration as a popular exhibition rather than the serious examination of the whole principle of the machine he craved. But the one person who really mattered, Lord Northampton, the President of the Royal Society, was so impressed that he invited Fowler to prepare a detailed description of the machine for their next full meeting on June 18th, just three weeks later.

Buoyed up by his success, Fowler set out on the return journey to Devon. Coaches left the Bull Inn, Aldgate, on Mondays, Wednesdays and Fridays at 2.30pm and passed through Great Torrington at 6pm on the following day en-route for Cornwall. Hounslow was the first stage for changing horses. In its heyday as many as three hundred coaches passed through each day before struggling across the heath on roads up to two feet deep in mud in the winter. These conditions were manna from heaven for the highwaymen, many of whom ended their days hanging in chains as a gruesome warning to others.

The gibbets were gone, but sadly so was all the hustle and bustle. The town was declining fast now the railway was open as far as Slough. Not only had hundreds of passengers deserted the stagecoaches, but also vast herds of cattle and sheep had been taken off the road and put into railway trucks. The fixtures and furnishing of the Inns were up for auction as once-prosperous innkeepers called it a day.

Newbury was the next stop, a town that now held much greater significance for Fowler than on his outward journey. This was the birthplace of Francis Baily. He was nicknamed the Philosopher of Newbury for his obvious intelligence but a love of adventure dominated his early life. In 1795, when he was just twenty-one, Baily embarked on an extraordinary journey to the United States. He survived two shipwrecks, a voyage in an open boat down the Ohio and Mississippi rivers from Pittsburgh to New Orleans, and a return journey to New York across nearly two thousand miles of wilderness.

His next venture was to be an exploration of the Niger, but funds ran short and he was persuaded into a career as a stockbroker. He amassed a modest fortune before retiring to pursue his real passion, astronomy. He was one of the founding members of the Astronomical Society, and gave his name to Baily's Beads, a phenomenon seen just before the moon covers the sun during an eclipse.

Above all, Baily was considered a just man, inspiring respect and affection. These qualities would have made a deep impression on Fowler when the two met at Hunter Street, and were the key to engaging Baily's cooperation at the Royal Society meeting just three weeks later.

Fowler must have been very satisfied. His new invention had been scrutinised and praised by the foremost mathematicians and scientists of the day. Significantly, he had been invited to prepare a paper for the next meeting of the Royal Society. Their archives are full of letters from aspiring inventors trying again and again to get their inventions noticed, most never succeeding. Yet Thomas Fowler had actually been invited, by the President no less, to bring his unique calculating machine to their attention. At sixty-two years of age the fellmonger's apprentice from Great Torrington in the County of Devonshire was finally poised to achieve the recognition his pioneering work deserved.

Chapter 7
[+ − +]

In which history is in danger
of repeating itself

Back home in Torrington, Fowler threw himself into the less heady but familiar activities of the town. Mary and Caroline had a list of printing and book-binding commissions to discuss, there were council meetings to catch up on and a backlog of appointments with customers at the bank. But everyone wanted to hear about his trip, particularly his sponsors. Their belief in Fowler had been vindicated. This country man had conquered London! Carried away with enthusiasm for their protégé, Trevor Wheler and Lord Clinton made plans to attend the next meeting of the Royal Society where Fowler's paper would be read.

So why was Fowler pacing the floorboards late into the night instead of getting on with the task in hand? All the years of solitary study and labour in his workshop, this is what it had been for. Yet he was paralysed. It was fourteen years since Mr. Rotch had suggested he go hang or drown himself to get rid of his troubles over the Thermosiphon. Fourteen years that felt like yesterday. Now Lord Northampton had specifically asked that he put together the best possible description of his machine. That meant details, drawings, everything he, and only he, knew. His nights were plagued by dreams of the wealthy Gentlemen Fellows of the Royal Society scrutinising every detail and recreating his machine in their own workshops. He had to find a way to safeguard his invention. But how?

Finally, he came to a decision. It was his use of the signed, or balanced, ternary numbering system was the significant innovation. He would put together a brief description of the machine but then focus on a detailed analysis of the underlying arithmetic, something guaranteed to prove his worth in the wider scientific community. A comprehensive table could take the place of drawings. He sat at his desk overlooking Torrington Square and began to write. "The machine consists of four essentially distinct parts…" He swiftly settled into the underlying arithmetic, producing several densely packed pages of figures.[12] Then, on June 9th, 1840, his sixty-third birthday, he signed off with the statement "In this way any Arithmetical Operation, whether simple or compound, may be accurately performed and the whole is applicable to Mechanical Operation with little expense or trouble." Job done. It was a paper that showcased his mathematical ability without revealing the detail of the machine. That could come later, once he had been credited with the theoretical breakthrough.

Did Fowler allow paranoia to cloud his judgment? Perhaps, but he had good reason to be cautious. History is littered with examples of stolen ideas and patent disputes. His idea was completely original, a rare and enviable occurrence amongst academics. Some members of the Royal Society might well have been tempted to rediscover Fowler's invention in their own workshops and publish before him.

His sponsors were less impressed. Thomas Fowler was an honourable man of unquestionable integrity. Yet he could be impossibly frustrating at times. Recognition for his life's work was in the balance; all he had to do was send Lord Northampton a detailed description. Instead, he had generated page after page of the most complicated arithmetic that would go right over the heads of many at the meeting. It amounted to self-sabotage!

Fortunately for Fowler, another of his sponsors came to the rescue. After a distinguished army career, Sir Trevor Wheler he settled into a life of politics and philanthropy in North Devon. He helped establish a small school in Little Torrington and, in 1851, followed Lord Rolle and Lord Clinton as Commanding Officer of the Royal North Devon Yeomanry. More importantly for Fowler, Sir Trevor Wheler was a well-respected figure in London society. He had taken an interest in Fowler's work for some time, but fortunately realised he needed to become more actively involved if the exasperating, but undeniably

---

[12] See appendix 6 for Fowler's paper

brilliant, Mr. Fowler was not to lose this once in a lifetime opportunity. Through Sir Trevor Wheler's intervention one person in particular now became crucial to Fowler's success — Augustus De Morgan.

De Morgan was a highly qualified and thoroughly likeable mathematician with a reputation for being entirely free from self-interest. A browse through De Morgan's papers also reveals an  inveterate doodler and a man with a refreshing approach to his subject. Letters exploring complex mathematical theorems often ended on a lighter note. In one on tidal observations, De Morgan concludes "I hope the tides continue to behave themselves like decent oscillations." And on another occasion "I hope all your family have escaped the epidemic which is raging here. Every living soul in this house has it or has had it, except the cat, about whom I believe nobody has inquired."

Augustus De Morgan

De Morgan agreed to help with Fowler's presentation. He had witnessed both demonstrations at Hunter Street so had a reasonable grasp of the machine and its operation, but he still felt he had to apologise for "such an account of it [the machine] as can be given without drawings." Two people making their way to London for the meeting on June 18th, Sir Trevor Wheler and Lord Clinton, would have been able to read between the lines.

The one person not on the road was Fowler. The Statutes of the Royal Society at that time explicitly stated "No stranger, excepting Foreign Ambassadors and Ministers, and other distinguished persons specially invited by the President, shall, on any account, be permitted to be present during the Meeting." Sir Trevor Wheler and Lord Clinton were obviously considered distinguished enough, but not the humble author of one of the papers to be read.

This was not Fowler's only frustration. Sir Trevor Wheler had convinced him that De Morgan was the perfect representative for him. As Professor of Mathematics at University College, London, a prolific writer for the Society for the Diffusion of Useful Knowledge and a respected teacher, he was the ideal advocate. But he was not a member of the Royal Society. When the meeting arrived at item 10 on the agenda, *Description of a calculating machine invented by Mr. Thomas Fowler of Torrington in Devonshire*, it was Francis Baily, the 'Gentleman Philosopher of Newbury', who rose to deliver De Morgan's script.

> Mr. Fowler had been employed by the Guardians of the poor of the district in which he lives to calculate the manner in which a given total assessment should be levied upon the differed parishes of the union, each parish being in proportion to a given sum.

This much we know, but he continued:

> Mr. Fowler, having drawn up and published some of the necessary tables, and having found his method convenient in practice, began to consider whether the mechanical part of the calculation might not be performed by machinery. The instrument, which he caused to be constructed by the workmanship of a country carpenter, though large and difficult to move, is easily used. The following is such an account of it as can be given without drawings, and Mr. Fowler is occupied in preparing one of a more detailed and complete kind.

The country carpenter mentioned here was, of course, Fowler himself.

> The machine consists of four essentially distinct parts. The first, second and third exhibit the multiplicand, multiplier, and product, or quotient, divisor and dividend, according as the question to be worked is one of multiplication or division. The fourth is a carrying apparatus which though at present detached, and employed to reduce the result to its simplest form after the main operation has been performed might without much

difficulty be attached to the multiplier or divisor, and work with it.

A description of the operation of the machine followed, a combination of Fowler's report and Augustus De Morgan's own observations from Hunter Street. However, he concluded:

> The instrument confessedly does not realise the notion of a calculating machine properly so called, since the necessity for using tables, both for converting the factors and reconverting the result, introduces both labor and risk of error. It is also felt that the preceding description is insufficient to give more than a mere glimpse of the principle and detail. Such as it is, however, the Society will perhaps not regret the bestowal of one moment of its attention, when it considers that the inventor will thereby not only feel honoured and gratified, but secure in the possession of the credit to which his ingenuity is entitled, and for which alone he has labored.[13]

The future of Fowler's calculating machine now rested in the hands of the Royal Society. Some papers were published in their Transactions. If Roget now included De Morgan's it would amount to an official sanction of Fowler's work. With one stroke he could achieve his goal.

Lord Clinton and Sir Trevor Wheler were easily able to assure Fowler that his ingenuity and invention had impressed the Gentlemen Fellows of the Royal Society. But his use of signed, or balanced, ternary had become a double-edged sword. His audience was impressed, but worried about the potential for error in converting between ternary and decimal. With Babbage's automatic mechanism in mind, their other reservation was the amount of knowledge and labour required to operate Fowler's machine.

It was too frustrating. Automating the mechanism would be a further advance but they had simply failed to grasp the fundamental significance of his choice of ternary. Not that this was a surprise. The Royal Society had such an inbuilt resistance to any form of applied science that an appreciation of his very practical application was always

---

[13] For the full text see Appendix 5

unlikely. It was an entrenched attitude. Even Plato condemned the work of a contemporary who constructed machines based on mathematical principles, considering such an application "a degradation of a noble intellectual exercise, reducing it to the low level of a craft fit only for mechanics and artisans."

All Fowler could do was wait. He had plenty of work to occupy him and there were other pleasant diversions, such as the needle match that took place on August Bank Holiday Monday. All life in Great Torrington was suspended, or so it seemed, as everyone gathered on horseback or piled into coaches and wagons to make the ten-mile journey to Northam Burrows. In this spectacular setting, within earshot of the waves crashing on to the beach at Westward Ho! the gentlemen of Torrington secured a glorious victory over their arch enemy, Bideford — at cricket.

The letter finally arrived, but the news it contained was devastating. The Council had decided not to publish any of De Morgan's paper under their sanction.

The Royal Society promoted itself as an organisation intent on stimulating discovery and furthering the dissemination of scientific truth. "A truly worthy Corporation, whose splendid honours were not to be gained by wealth or by power, but by scientific merit alone, bestowed as the reward for honest study, high attainments and genuine work." Or so the publicity went. Fowler begged to differ. A brief report on his machine was included in the Proceedings but that was all.

Having come so far this was an agonising setback. They had not even had the courtesy to explain their action. Fowler's feelings are clear from a later letter to Francis Baily.

I have for a long time intended writing to thank you for presenting to the Royal Society some papers of mine transmitted to A De Morgan by Sir Trevor Wheler and was rather surprised to find a note from this gentleman to Sir Trevor saying that he had received a pro-forma notice from the Council (or something to that effect) that no part of them would be published under its sanction, this was indeed nipping things in the bud as the papers were open to any alteration or improvement, but I had no desire for this honor, until the Marquess of Northampton, in your presence, advised me to draw up the best description of the Machine in my power on my return to Torrington and send it up, which I did and I believe it was accompanied with a very lucid and able report by A De Morgan, Esq. which I have since seen and much approve.

I have no doubt but the council thought the whole unworthy of notice as the production of an humble individual like myself, but I may be allowed in extenuation of my temerity, to remark, that however humble a first idea may be it often leads to the most valuable and refined results, and the very circumstance of giving it honorable publicity is the means of such further improvements as could not at first be anticipated, conclusions such as these may be drawn from the whole history of science, which is only a series of successive improvements on simple and original ideas.[14]

Fowler had a talent for self-expression, choosing words and phrasing in this last paragraph that are positively poetic. But was he right? Had his work been rejected because he was a self-taught countryman? Had his hopes for academic credibility been sabotaged before he even set foot in London because he lacked the right connections?

A century earlier, in the summer of 1730 another self-taught inventor made the long journey to London from his rural home. As a young boy, John Harrison also became a skilled woodworker while working alongside his father, a country carpenter. A few years later, Harrison completed his first pendulum clock, acquiring a well-deserved reputation for accuracy. At the end of a month his clocks were less than a second out while the finest watches elsewhere consistently lost around a minute a day. In his simple country workshop, Harrison had outperformed the biggest names in clock-making in London.

Like Fowler, Harrison sought recognition for his achievement, setting his sights on the Longitude prize. In the early 1700s too many ships were lost at sea because they were unable to determine their exact location. Parliament's solution was to announce a reward for anyone who could solve this problem. To stake his claim, Harrison travelled to

---

[14] See Appendix 11

the Royal Observatory in London. Dr Edmond Halley, the Astronomer Royal, viewed Harrison's work with an open mind — others resisted any thoughts of a mechanical solution to what was seen as an astronomical problem — referring him to George Graham, a prominent watchmaker and a member of the Royal Society.

Harrison immediately faced the problem all too familiar to Fowler: how to gain recognition without losing the credit? Harrison was lucky, George Graham proved a trustworthy patron and enthusiastic sponsor. With his recommendation, the Board of Longitude agreed to fund Harrison's research. He was even offered a Fellowship of the Royal Society, recognition Fowler could only dream of. There are many parallels between the life of John Harrison and that of Thomas Fowler but, initially at least, Harrison succeeded despite his humble origins.

A century later, but on a similar quest, was Fowler right to claim discrimination because of his background? Academic snobbery was rife and without a university education, membership of this elite club was hard won. But perhaps he overstated the case, deflecting the blame from his refusal to supply drawings. Where John Harrison had George Graham to promote his work, Fowler had Augustus De Morgan. But Graham had an intimate knowledge of Harrison's mechanism. Indeed the clock had been left with him to present to the Royal Society. De Morgan, on the other hand, could only share his observations from the two demonstrations and the theoretical background in Fowler's paper. He was left apologising for being unable to give more than a brief outline of the calculating machine. Was this simply not enough to justify inclusion in the Transactions, however potentially significant the subject matter?

Whatever the cause, The Royal Society's decision was devastating. Archdeacon Stevens and Sir Trevor Wheler had lifted Fowler from his provincial existence into the society of mathematicians and scientists who admired his work. It was a lifelong ambition fulfilled — almost. Momentary recognition was not enough. He still wanted a thorough examination of the whole principle of the machine, for it to be universally accepted and for his part to be properly acknowledged. Perhaps they had viewed him as a temporary distraction after all rather than a man of science with a device of great potential significance.

Fortunately the Royal Society was not the only scientific body in town. In 1831 a small group of academics, with Charles Babbage at the helm, became so frustrated with the increasing conservatism of the Royal Society that they formed the British Association for the Advancement of Science; the BAAS. Babbage hoped the Association would influence government policy and support new research, particularly in applied science. The Association grew in popularity, reaching its peak as a forum for serious scientific debate in the late 1830s with hundreds of participants converging on a different venue each year for their annual meeting.

Dickens was quick to parody these nomadic assemblies in his fictional reports of the Mudfog Association for the Advancement of Everything. His bulletins speak of locals in the host town displaying "a wildness in their eyes, and an unwonted rigidity in the muscles of their countenances, which shows to the observant spectator that their expectations are strained to the very utmost pitch." Dickens extends his irreverent view with characters delivering papers on their quirky research. One of the actual papers on the agenda at the 10th annual meeting of the British Association being held in Glasgow in 1840 may well have been mistaken for a Mudfog entry. Professor Brewster was speaking on "The Cause of the Increase of Colour by the Inversion of the Head." Not, as you might think, an explanation of why people go red in the face when bending down but "an examination of why the colours of scenery are augmented by viewing them with the head bent down and looking backwards between the feet. Blue and purple tints of distant mountain scenery are thus beautifully developed."

Parody aside, the Association was a valuable forum for new ideas. Mathematical papers were heard by Section A, and subsequently covered in *The Athenaeum*, a London journal. The next time Fowler heard from his patron, Sir Trevor Wheler, it was to draw his attention to the issue for October 31st, 1840. Here he found detailed accounts of scientific progress given by many familiar names, including the Astronomer Royal, Professor George Bidell Airy. Only this time he was not reporting on his own work, but Fowler's.

The origin of this machine was to facilitate the Guardians of a Poor Law District in Devonshire in calculating the proportions in which the several divisions were to be assessed. The chief peculiarity of the machine was, that instead of our common decimal notation of number, in it a ternary notation was used; the digit becoming not tenfold but threefold more valuable as they were placed to the left.[15]

Even better, the Association included a report in their Transactions.[16] Despite failing to attend either of Fowler's demonstrations, Airy had somehow been persuaded to review De Morgan's paper at the British Association.

Within four months, Fowler's invention had been brought to the attention of two of the foremost scientific bodies in the land. He had received a positive reception from the Fellows who witnessed his demonstrations, and, thanks to this review, details of his innovative calculating machine were now well and truly in the public arena. Limited success compared to that of John Harrison, but still a satisfying confirmation of the significance of his work.

However, after initial success, both Harrison and Fowler suddenly found wider recognition of their work brought to an abrupt halt by individuals occupying the same pivotal role — that of Astronomer Royal.

After taking almost twenty years to develop his third clock, Harrison discovered his relationship with the Board of Longitude changed by the arrival of a new Astronomer Royal, Dr James Bradley, and a future Astronomer Royal, Nevil Maskelyne. Harrison suffered frustrating delays as the goal posts were constantly moved to favour an astronomical rather than a mechanical solution to the Longitude problem.

In 1761, when Harrison's fourth, and radically different time piece, was finally tested on a voyage to Jamaica, its performance was outstanding, losing only five seconds after eighty-one days at sea. His watch had done everything required of it by the Board of Longitude. The prize should have been his. Instead, more obstacles were placed in his way: better verification, new trials, more instruments.

---

[15] ©The British Library 251.1.16-20
[16] See appendix 7

It was three long years before the Board of Longitude reluctantly conceded that Harrison's watch had performed well within the requirements of the Act. But he was still only offered half the prize and it took the intervention of King George III for this travesty to be brought to an end. Harrison was eventually granted the full amount three years before his death in 1776 — a few months before Fowler was born.

In 1811 the post of Astronomer Royal passed from Nevil Maskelyne to John Pond and then, in 1835, to George Bidell Airy. Airy was born in 1801 in Alnwick, Northumberland. As a young boy, he performed well at school but was not liked by his schoolmates — a sentiment that appears to have been mutual. He made some gains in the popularity stakes through his skill in constructing peashooters but he was considered a snob. While still young, Airy went to live with his uncle, Arthur Bidell, an educated man whose acquaintance with many leading scientists helped equip Airy for a place at Trinity College, Cambridge. Airy graduated as the top First Class student in 1823, although some thought this had more to do with his exceptional memory and organisational skill than any mathematical brilliance. He verged on the obsessional, keeping a record of every thought and communication from the time he went up to Cambridge to the end of his life.

Three years after graduating he was appointed Lucasian Professor of Mathematics at Cambridge, the first in a string of academic appointments, and of several triumphs over Babbage. The Board of Longitude came next — also sought by Babbage — followed by other posts until, in 1835, he became Astronomer Royal. Airy acquired a reputation both as a hard taskmaster and an extremely hard working man himself.

George Bidell Airy

He published over five hundred papers and reports and eleven books, some so complex they were considered unreadable by anyone not already an expert. He was a heavyweight in the worlds of mathematics and astronomy and, significantly for Fowler, massively influential in the evaluation of new inventions.

After reviewing De Morgan's paper Airy was sufficiently interested to forward both the paper and Fowler's accompanying Tables to his colleague, Professor Phillips. But his frustration is clear in this note to another colleague, Professor Forbes.

> Royal Observatory Greenwich
> My Dear Sir
>   I have just dispatched to Professor Phillips, St Mary's Lodge, York, the account of the Calculating Machine for the ternary system and the accompanying Tables. I would have written a few lines of account of it, but, after reading it again, I have no clearer idea of the details that I had — It is quite wrong to send a description of machinery without drawings.

Apparently Airy was unable to visualise the machine and its operation from the glimpse De Morgan's paper offered. Despite being invited, Airy had not attended either of the demonstrations given by Fowler at Hunter Street earlier in the year so suffered more than anyone from the absence of clear plans. Fowler's fear that he might miss out on any acclaim for his invention was swiftly becoming a self-fulfilling prophecy as all went quiet through the winter months of 1840.

Chapter 8

[+ 0 −]

The Third Law

Thomas Fowler and John Harrison were not the only inventors to suffer at the hands of an Astronomer Royal.

In August 1840, Charles Babbage accepted an invitation to speak on his Analytical Engine at a meeting of Italian scientists in Turin. Babbage revelled in the admiration expressed not only by these scientists, but also their King. But their enthusiasm failed to translate into an official report that might have swayed government opinion at home. Sir Robert Peel was now Prime Minister, a conservative man who used his office to install like-minded individuals in prominent positions. When George Bidell Airy was appointed advisor to the government on scientific matters, many believed he acquired an influence in excess of his ability. Crucially, it was Airy that the Chancellor of the Exchequer consulted on the subject of continued funding for Babbage's calculating engines.

When the Royal Society initially recommended government support for Babbage's work on the Difference Engine they commented "The Committee cannot but observe that, had inferior workmanship been resorted to, such is the number and complexity of the parts, and such the manner in which they are fitted together, the success of the undertaking would have been hazarded; and they regard as extremely judicious, although, of course, very expensive, Mr. Babbage's determination to admit of nothing but the very best and most finished work in every part; a contrary course would have been false economy, and might have led to the loss of the whole capital expended on it."

As a result Babbage received his first grant and work on the Difference Engine began. Over ten years and £17,000 of public money later [around £1,359,000 today] Airy presented a very different picture. Regarding the Royal Society's comments Airy stated "These persons were all private friends and admirers of Mr. Babbage and, without laying any things to their charge which could not be ascribed to the most honourable man living, I cannot help thinking that they were a little blinded by the ingenuity of their friend's invention … When the Report was discussed by the Council of the Royal Society, it was boldly stated by Dr Young … that, if finished it would be useless."

Airy continues to berate both Babbage and his invention. "Mr. Babbage made the approval of the machine a personal question. In consequence of this, I, and I believe other persons, have carefully abstained for several years from alluding to it in his presence. I think it likely that he lives in a sort of dream as to its utility." And finally pens his unequivocal conclusion. "I can therefore state without the least hesitation that I believe the machine to be useless, and that the sooner it is abandoned the better it will be for all parties." Airy was not a man to sit on the fence.

In November, 1842, Babbage received a letter confirming that the Government, on the grounds of expense, was abandoning further work on the Difference Engine. This expense had been considerable. In addition to the public funds made available to him, Babbage personally contributed around £20,000 in cash or in kind for which he had a favourite anecdote.

A short time after the arrival of Count Strzelecki in England, I had the pleasure of meeting him at the table of a common friend. Many inquiries were made relative to his residence in China … Count Strzelecki told them that the subject of most frequent enquiry was Babbage's Calculating machine. On being further asked as to the nature of the enquiries, he said they were most anxious to know whether it would go into the pocket … I told the Count that he might safely assure his friends in the Celestial Empire that it was in every sense of the word an out-of-pocket machine.

Thirty-five years later, the British Association for the Advancement of Science re-assessed Babbage's work. Despite considering that "the labours of Mr. Babbage, firstly on his Difference Engine, and secondly on his Analytical Engine, are a marvel of mechanical ingenuity and resource" they decided against continuing to support his work.

Their report was forwarded to Airy who evidently had not changed his opinion over the years. "Though I have been pleased — from a very distant time — to make myself acquainted with the principles of those engines, I have never thought that they deserved expense, because they could be used so very rarely." Airy was a brilliant mathematician but blinkered to the potential of mechanical devices in the scientific world. It was a popular view reflecting the developing split between pure academic science and the so-called inferior discipline of applied science. He saw no potential in Babbage's work and had no qualms about settling the matter as he saw fit.

It was against this background that Augustus De Morgan's paper on Fowler's mechanical calculating machine arrived on Airy's desk in the summer of 1840. He was being presented with yet another mechanical device of dubious utility, and without any plans to make the mechanism understandable. He was intrigued by the principle behind the machine, but tragically, this was not enough to counteract his inbuilt prejudice.

The winter of 1840-1841 must have been a depressing time for Fowler. After all the excitement of London where was he now? A few months ago he had not only been in the same room as the great and good of the scientific world — Faraday, Baily, Beaufort, De Morgan, Northampton — but they had shaken him by the hand and praised his machine. No less a person than Charles Babbage had admired his work. He was proud of what he had achieved in his various roles within Great Torrington but this was the company he longed for.

His response was to retreat to the familiar surroundings of his workshop and a problem he could deal with — improving his machine. One thing was obvious: he needed to simplify the conversion from decimal to ternary, and vice versa, before exhibiting it again. But time was short. The next AGM of the British Association was in Plymouth, just a day's journey from Torrington, and Sir Thomas Dyke Acland, an acquaintance of Sir Trevor Wheler, had been appointed Vice President. It was too good an opportunity to miss.

Sir Trevor Wheler was looking a little less further ahead. He had received a copy of Airy's note where he complained that "it was quite wrong of Fowler to send a description of machinery without drawings". Fowler really was his own worst enemy. He had ignored Lord Northampton's request to supply the best possible description of the operation of the machine, and now this. But the note provided an obvious solution. Fowler simply had to supply Professor Airy with the plans and sections that would make the mechanism understandable. However, Fowler's paranoia, or judicious caution, had become more, not less, entrenched over the years. Flouting what amounted to a direct instruction from his social superior and patron, he refused.

There must have been something very charismatic about Thomas Fowler. Despite undermining the efforts of his supporters at every turn, they stuck with him. Even with one hand now firmly tied behind his back, Sir Trevor Wheler tried to re-open doors with a letter to Airy:

> Sir … I was in hopes Mr. Fowler might have been able to furnish some Plans & Sections of his Machine, but he tells me he is not prepared to offer any further details of its construction than those which have already appeared in a Paper drawn up by Mr. De Morgan and read, I believe, at the meeting of the British Association last year.
>
> He has however, requested me to forward the enclosed letter and I venture to accompany it with an attempt to explain in some degrees the operation of the instrument by shewing the manner in which the Addition, Subtraction and Multiplication of the Signs used by Mr. Fowler to express his numbers is performed on paper, since it represents exactly the process that takes place in the Machine itself; for nothing can be more simple than its construction.[17]

Sir Trevor Wheler expanded on the underlying mathematical principles in the hope this would address Airy's frustration. But other than this, he could do no more than forward Fowler's letter.

---

[17] See appendix 8

My friends and patrons here, among whom I would particularly name Sir Trevor Wheler, are anxious that I should thank you for your notice of my new calculating machine at the last meeting of the British Association, as reported in the Athenaeum of 31 October last. Please to accept my sincere and hearty thanks for this honour and if I may be permitted to intrude on your valuable time, I would give some further account of this machine, which perhaps may hereafter be found valuable as well in small as in the most extensive arithmetical calculations …

I can easily understand your feelings at the (perhaps) unexpected result of a differential equation unravelled entirely by your own patience and ingenuity; such possibly, though of a lower grade; are any feelings of the results of my mechanical operations; conscious that the whole is as yet untrodden ground I would in illustration give one instance; — since I left London I have applied the Machine occasionally to the calculation of Logarithms, and, from the comparative ease with which this may be done even to 70 or 80 or indeed to any number of places of figure, and, with the same facility either for the Logarithm or Natural Number, I am almost inclined to think that this may properly be called a Logarithmic Machine. My present clumsy performance will give a Logarithm or Natural Number true to 26 or 27 places and the limit is only the extent of human mechanical agency.

Fowler continues to expand on the capacity of his machine, going into greater detail on the Ternary system and commenting that:

I often reflect that had the Ternary instead of the Denary Notation been adopted in the infancy of Society, Machines something like the present would long ere this have been common, as the transition from mental to mechanical calculation would have been so very obvious and simple.[18]

This simple, unassuming statement holds the key to the fundamental significance of Thomas Fowler's visionary work. Through his choice of

---

[18] See Appendix 9

a low-order numbering system he opened the door to simpler calculation, both in his tables and mechanically, a leap of imagination re-discovered by the early computer pioneers almost a century later.

He also produced a fully functioning machine that could have been constructed at a fraction of the cost of Babbage's Difference Engine. The significant words here are 'could have'. Sadly both Fowler and Babbage displayed a remarkable stubbornness in dealing with significant others; Babbage in his communication with the government and Fowler in his letter to Airy. Both clung to their point of view, refusing to grasp the opportunity to move forward on someone else's terms. Having been given a second chance with Airy, Fowler concludes "I am very sorry that I cannot furnish any drawings of the Machine."

It seems Fowler was determined to sabotage his sponsor's efforts to re-establish contact with the one person who could recommend government backing for his work. Instead of conceding the need to compromise, he requested what amounts to a favour:

> I hope I shall be able to exhibit it [the machine] before the British Association at Devonport in August next where I venture to hope and believe I may again be favoured with your valuable assistance to bring it into notice — I have led a very retired life in the Town without the advantage of any hints or assistance from any one, and I should be lost amidst the Crowd of Learned and distinguished Persons assembled at the Meeting, without some kind Friend to take me by the hand and protect me.

Fowler obviously felt his isolation from academic life deeply but was he really as out-of-his-depth in these gatherings as he claims? The dogged determination, resilience, and self-promotion evident in his life within Torrington suggests otherwise. Perhaps this attempt at ingratiating himself was simply Fowler trying to mitigate any irritation caused by his intransigence over the drawings — or lack of them. He did not have to wait long for a response. Airy's reply to Sir Trevor Wheler began:

> Sir. I beg to acknowledge your note of the 13th inst. inclosing a letter from Mr. Thomas Fowler. The copy of my letter to Professor Forbes to which you allude was, I suppose, the copy

of a note in which I explained why Mr. De Morgan's account of Mr. Fowler's machine was not sufficiently complete for my understanding. I am much assisted by your explanation of the process of subtraction, in regard to one point; namely the advantage of using the symbols 0, +, −, instead of 0, 1, 2; but with regard to the general construction of the machine I am still in great obscurity.

I may mention as a general remark that I have little belief in the extensive utility of machines of this class. The number of persons who can use even the common sliding rule is very small. I do not mention this as tending to exclude absolutely the advantage of such mechanical contrivances, but as tending to limit it greatly. They will only be used where there is a systematic education, thus all my Assistants are instructed to use the Sliding Rule and I believe that officers of the Excise are trained to use it before they undertake active duties. Without this express training the sliding rule would not be used, This applies more strongly to a more complicated machine especially when the previous instruction is so heavy as it must be in Mr. Fowler's.

You will have the goodness to understand these remarks as in no degree delegating from the ingenuity of Mr. Fowler's contrivance, which I believe is very great; nor from its utility as used by himself or persons immediately around him; but only as expressing my views as to the extent of its utility to other persons.

I beg to acknowledge your kindness in inviting me to call on you at Cross House. I believe however, that I shall not attend the meeting of the British Association in the present year.

I am, Sir, Your very obedient servant, G.B.Airy[19]

It is interesting that Airy praises the ingenuity of Fowler's machine and its potential when used by him. But yet again, Airy was going to miss

---

[19] By permission of the Science and Technology Facilities Council and the Syndics of Cambridge University Library. RGO60/4 Misc. Correspondence 1828-1843

Fowler's demonstration of his machine, and as far as he was concerned this was the end of the matter.

In 1990 F.J.M.Laver, then President of the Devonshire Association, provided a contemporary reflection on Airy. "[Babbage] suffered also because the then Astronomer Royal, who really should have known better, completely failed to perceive the usefulness of such a machine, and put in a spiteful report to the government which cut off all further funding. In so doing he provided a paradigm example of Arthur Clarke's Third Law, which states that when an eminent elderly scientist declares that something or other can be done he is probably right; but when he says it cannot possibly be done he is almost certainly wrong — and liable to trip over a younger man who is already doing it."[20] The irony here, of course, is that both Babbage and Fowler were actually older than Airy and perhaps, in this respect at least, wiser?

Some years later Airy had good cause to reflect on his opposition to the Difference Engine. In 1857 tables published at government expense and under his direction contained one hundred and fifty-five errata. Then, incredibly, the errata were found to contain mistakes and an errata of erratum was published. Eventually the list of errata was larger than the original tables! Babbage commented "In making these remarks I have no intention of imputing the slightest blame to the Astronomer Royal, who, like other men, cannot avoid submitting to inevitable fate. The only circumstance which is really extraordinary is that, when it was demonstrated that all tables are capable of being computed by machinery, and even when a machine existed which computed certain tables, that the Astronomer Royal did not become the most enthusiastic supporter of an instrument which could render such invaluable service to his own science."

---

[20] Rep. Trans.Devon.Ass.Advmt.Sci., 122 1-16

Chapter 9
[+ 0 0]

Fowler's Universal Calculator

In July 1841, the British Association for the Advancement of Science took its annual meeting to pastures new in Plymouth in the south-west of England. In earlier years this sea-faring town was the gateway to an ever-expanding world. Captain James Cook and Sir Francis Drake departed from here, returning with tales of new worlds and more than a little plundered treasure. Drake was swiftly elevated to national hero after becoming the first Englishman to circumnavigate the globe, though he is more popularly remembered for finishing a game of bowls while the Spanish Armada advanced.

More than two centuries later Plymouth braced itself for an invasion of a very different kind, as members of the British Association arrived from across the country. Many of the four hundred and fifty who attended were already significant in Thomas Fowler's story; men such as William Buckland, the Marquess of Northampton and Francis Baily.

It is to Fowler's credit that he was also on the road to Plymouth. A lesser man might have decided enough was enough. In many ways he had achieved his aim. His machine had been praised by the finest scientific minds of his time, and even Airy accepted its value when used by Fowler. So after the demonstrations in London why did he not return to Great Torrington, refine his machine and sell it commercially? What convinced him that some of his country's most respected mathematicians had got it wrong? Misguided ambition, arrogance or sheer stubbornness all qualify. However, Fowler's passionate belief that the full significance of his invention had yet to be realised is a more likely candidate.

At the end of July 1841, Fowler again painstakingly packed his machine into a wagon to protect it from the rigours of the road and set off for Plymouth. By late afternoon on Monday 2nd August he had assembled it at the Naval Annuity Centre in Ker Street, Devonport, and checked and re-checked that it was functioning perfectly.

Early next morning members of the Association dispersed to separate venues to hear papers on various topics. The Geology section considered Dr. Buckland's considered opinion that Chichester in Sussex was more liable to earthquakes than any other place in the country, while those in Section A, Mathematics and Physics, gathered in the Hall of Athenaeum.

Inside the Hall of Athenaeum.

There were nineteen items on the agenda, item 17 being *Mr. Fowler on a Calculating Machine*. Interestingly, item 16 was on Professor Moseley's Calculating Machine. Henry Moseley, then Professor of Mechanics at King's College, London, also tried to make contact with

Airy, with even less success than Fowler. His paper on a machine *For calculating the products, quotients, logarithms and powers of numbers* was returned unread. Airy wrote "Mr. Moseley's paper has been lying on my table for a month. I could not look at it at first and I have no chance of doing so now. I have therefore returned it." At least Airy read De Morgan's paper.

For years, Fowler had laboured day and night, investing everything in his invention. The Royal Society's decision not to publish was a huge disappointment but he was determined he would not be side-lined again. Surely the scientists gathered in Plymouth would see, as he did, that machines like his would have been in use years earlier if the principle he had applied had been adopted sooner. All he needed was for this demonstration to be successful for the history of calculation to take a very different route.

But as Fowler stood to deliver his presentation at the Hall of Athenaeum, his machine stood in splendid isolation at the Naval Annuity offices in Ker Street, a completely different location. Regardless of whether it was a mistake by the British Association or a misunderstanding on Fowler's part, he was unable to demonstrate his machine as an integral part of his talk. It was devastating. All the preparation, the hours spent ensuring his presentation would be flawless — all for nothing. He could talk about his machine but as Airy and Northampton had made all too clear, without drawings or the machine itself it was pointless.

An arrangement was hastily made for interested members to attend a demonstration the following day, but this was after the official end of the meeting. Instead of being able to take advantage of a captive audience, Fowler was now faced with the prospect of persuading delegates that it was worth putting themselves to considerable trouble to inspect his machine.

To have come so close and be thwarted yet again was almost a setback too far. His sponsors looked for the positives. At least *The Athenaeum* included a report on his machine and The British Association *Transactions* carried a fuller description:

> The machine itself is on the principle of the old abacus, or calculating rods. At first glance it has somewhat the appearance

of a pianoforte, or organ, with all its keys laid bare. It would be impossible by mere description to make clear the details of the instrument, or the method of working it. The mere mechanism, although ingenious, is not, we think, the best suited to rapid movement, nor altogether free from clumsiness of execution: this, we understand, has arisen from Mr. Fowler having made every part of the machine with his own hands; the assistance of workmen having been avoided through fear of piracy. What we most admired was the arithmetic on which the notation of this machine depends.[21]

"What we most admired was the arithmetic on which the notation of this machine depends." Music to Fowler's ears, and the feature that makes his work so relevant today. But he was frustrated at their focus on the clumsiness of the machine. Surely it was obvious that when he had been fully credited with the underlying principle and awarded funding he could build a far superior metal version. Concentrating on the arithmetic was the important thing — but instead the review moved on to a depressingly familiar reservation. "The great defect of it at present is that the translation of a given number into the ternary combination of signs suited to express it requires the aid of very voluminous tables. We can conceive however, that some simple method of translation may be devised, and to this we beg to direct the attention of its ingenious author."

Ingenious: clever, resourceful, original, inventive, creative, inspired, imaginative — all very appropriate definitions provided by a quick glance at Roget's Thesaurus. But the family were probably still treated to an outburst of indignation from Fowler. What was wrong with Tables? And even if the writer had a point, his signed, or balanced, ternary notation created visual patterns on the page, making errors much easier to spot than with a conventional table. Maybe the problem was that his invention was just too simple. His audiences were used to the complexity and sophistication of Babbage's gleaming machine: a clumsy wooden construction hauled from the provinces by an elderly man with a broad Devon accent was always going to struggle for their attention.

---

[21] See Appendix 10 for the full report

The question now was whether Fowler could set aside his indignation and focus on the positives. The review pointed the way forward. "We can conceive however, that some simple method of translation may be devised, and to this we beg to direct the attention of its ingenious author." Would he grasp this opportunity or remain stubbornly intransigent as he had with Airy? Or perhaps this time he would decide to call it a day.

The challenge to find a 'simple method of translation' gripped the imagination of one reader. "I read with much interest the account given of Mr. Fowler's calculating machine in a late number of the Athenaeum (p.700), and could not but be struck with the justice of the observation there made, that the most remarkable feature of Mr. Fowler's contrivance is the system of numerical notation adopted in it." The correspondent, who simply signs himself W.D.C., shares his opinion that "the transference of the ordinary numbers into Ternary ought to be effected by the machine itself."

Sixteen months on from the first demonstration of his machine it would have been all too easy for Fowler to give up. He was sixty-four. He had expended an enormous amount of time, energy and expense trying to gain recognition for his work. Even though the outcome had fallen short of his expectations, he could take comfort from the fact that many of the most respected scientists of the day were now aware of Thomas Fowler, the inventor from Great Torrington.

Meanwhile his responsibilities at home were as heavy, if not heavier, than ever. As well as the demands of the printing shop, the bank, the council, the Poor Law and the church, he was handling all the finances for a new Pannier Market being built at the end of the square. And there was little respite at home. His three elder boys were all away at school or university but they were still a family of eight occupying the rooms above the business. Caroline and Cecilia were working full-time in the print shop, with Frances, now fourteen, also helping out. But he had the three younger boys — Newell, eleven, and the twins Paul and Silas, just seven — under his feet. It would have been more than understandable for Fowler to focus on his business, family and community roles and head for a comfortable and contented retirement.

Understandable perhaps, but not an option for Thomas Fowler. He was about to demonstrate a remarkable resilience in the face of constant set-backs.

Throughout the autumn of 1841 he again shut himself away in his workshop determined to provide his critics with the simpler method of translation between decimal and ternary they were insisting on — a task that should be easily achievable by Christmas. But returning to the machine with fresh eyes was the trigger for a startling revelation. All thoughts of tweaking the ternary/decimal conversion were forgotten when it dawned on Fowler that even he had underplayed the significance of his invention. This was not just a ternary machine! By taking such a fundamental leap away from the design of all previous calculating devices he had opened the door to a machine capable of operating in any scale. Binary, ternary, decimal, in fact anything up to radix 30. It was an astonishing discovery, and it was his innovation of sliding rods in place of rotating wheels that made it possible.

In Babbage's decimal machine, each wheel rotates ten places. It is impossible to change to any other number system without taking the machine to pieces and starting again with new wheels. Carrying numbers forward also has to take place during a calculation, an impressive sight as they ripple along the machine, but creating still more limitations. Fowler's combination of sliding rods, and a carry mechanism employed after the calculation was more or less complete, avoided all this mechanical complexity. By simply inserting additional rods he could change the number system at will.

This discovery was too exciting, too significant to keep to himself, but who could he share it with? It was obvious that Airy was a lost cause, but Babbage, Buckland or Northampton might have been sympathetic. In the end, he decided to place his trust in Francis Baily, the Gentleman Philosopher of Newbury. In October, 1841 he wrote:

> I have the whole construction of such a Machine in my mind's eye and could give a rigid demonstration of its practicability before a file or hammer be used, this Machine would include and might be used for any Scale from Radix 30 down to the Binary and all the machinery of those scales would in a manner be in operation in a calculation by this splendid machine.

Fowler identified the only limitation for this new machine as being 'the extent of human capability for the mechanical construction.' He had a point. The scale of 30 would make considerable demands on construction. Rods might well be nudged along so far by the repeated actions of the machine that they ended up on the floor! But all these considerations would have been nothing more than minor hiccups once his machine was reduced in size as a result of, as he points out 'accurate workmanship such as might easily be procured in London.'

This vision was all it took for any thoughts of retirement to be banished. Fowler was on the scent again with as much fervour as ever; but this time tempered with pragmatism. His second machine had such potential, but for now Fowler knew that he had to prove the worth of his invention in the language everyone was familiar with — decimal. Retreating from the mechanism he now believed possible, Fowler continued his letter to Baily:

> … but my chief object at present being confined to the common decimal scale, I have made a model in wood of one term suiting this scale just as I should employ it in the Machine and I find its action perfect, I am therefore certain that the principles of the Machine you have already so kindly noticed, can be fully adapted to the denary notation at once, and that by it, all Arithmetical operations can be performed mechanically with great dispatch and certainty and the results instantly be presented if required as they are produced and stand in the Machine."

Sir Trevor Wheler was first to share the news locally. "Mr. Fowler is occupied in making improvements in his calculating machine by which he can now use the Decimal instead of the Ternary scale, and therefore it is more likely to be brought into general use. I wish the idea had suggested itself to him previous to the Plymouth meeting, as then he could have met all the objections raised to his system."

But the idea had not occurred to Fowler earlier, fortunately many might say. In creating a machine capable of operating across all these scales, Fowler was in danger of diluting the very thing that made his machine so significant — optimum calculating efficiency. And it inevitably meant abandoning the beautifully simple notation of −, 0 and + used in his ternary machine.

However, these doubts underestimate Thomas Fowler. He was already working on a new and equally simple solution, as he explained to Baily. "If the Scale of 30 were in operation I have invented a very simple notation for the digits of this scale which consists of the 10 characters of common arithmetic and a few letters of the Alphabet."

Throughout his life, Fowler experienced many seminal moments, from his first sight of *Wards Mathematicks* to the times he 'hit on an idea' with the Thermosiphon and his calculating machine. But it was with a new zest for life that he now returned to his workshop, determined to complete a greatly improved second machine that would not only answer, but exceed all the doubts and questions posed by his critics. Surely he would now achieve the recognition he craved.

But, despite the demonstrations he had given to so many eminent scientists, he still lacked someone to take up his cause. He decided to turn again to Francis Baily to help him over this final hurdle. His letter continues:

> "I hope, Sir that you may not consider this letter an impertinent intrusion, I have long looked on you as the best friend I could apply to, whose station and ability were sufficient to bring this matter into proper notice, and if the whole or any part of what I have now written after proper correction, be published in the Times and other papers under the influence of your name, it will draw attention to the subject, and I am perfectly free and willing to do any thing that may be in consequence required."

Despite this protestation, it is obvious that thirteen years on, his experiences with the Thermosiphon still haunted him. The familiar devil had to be addressed. He continues "My chief object in now writing to you is to beg that you will protect my claim to this invention." But his work was in danger of being superseded. Other devices were gaining publicity:

> "I should not have yet written to you in all probability had I not seen in the Times Newspaper of the 13th instant an account of a new calculating machine by Dr Roth of Paris. This may be a valuable instrument or only a Toy, at any rate I now feel that I ought to be no longer silent in regard to my own invention as I

am fully confident of producing a Universal Calculating Machine, a desideratum which has been so much sought after for many years past and of which the Machine I exhibited in London last year is the very first rudiment and foundation.[22]

Fowler's confidence in his second machine leaps off the page. He had found the solution to a problem so many had worked on for so long. He had to grasp the moment, particularly given the publicity for Roth's machine by *The Mechanics Magazine*:[23]

A new calculating machine has been invented by a Dr Roth of Paris, and presented in this country by a Mr. Wertheimer, which, the TIMES states, performs sums in addition, subtraction, multiplication and division with unerring exactness. Mr. Babbage, our own great experimenter in this line, is stated to have expressed his most unqualified admiration of this novel machine.

This review must have been particularly galling. Given Babbage's response, Fowler had every reason to believe that Roth's calculator was a real threat to his own. But his choice of Francis Baily as advocate was unfortunate. Earlier that year, Baily had been involved in a road accident and laid unconscious for a week. He was well on the way to a complete recovery by the time he received Fowler's letter, but having lost so much time he was now absorbed in his own experiments.

Perhaps if Thomas Fowler had been younger, and the entrepreneur he evidently was when he established himself as a printer, he might have opted to promote his machine in the commercial rather than the academic world. The Industrial Revolution was a time of unprecedented innovation and progress, when people from modest beginnings could acquire wealth and status through inventing for commercial gain. Talented entrepreneurs emerged, such as Richard Arkwright, the inventor of the power loom. Arkwright not only had the necessary skills and ability to invent new processes, but the drive to implement his ideas. Companies were beginning to realise that new

---

[22] See Appendix 11
[23] ©The British Library RB23.b.6538

approaches to the selling of products were as important, if not more so, than the technology itself. The Wedgwood group carved out a niche with products designed to appeal to a specific audience. Market research had arrived. Within this climate of opportunity, it was the innovators with the personal resources and financial backing to pursue their vision in a commercial manner who achieved success.

Fowler had limited personal resources and no government backing, but even with these, it is unlikely he would have been swayed from the path he had chosen. Like a dog with a bone, he remained focused on the academic world for the recognition he craved.

Chapter 10
[+ 0 +]

A Common Fate

Thomas Fowler's tenacity in the face of so many obstacles was remarkable, but there are some things that willpower alone cannot overcome. Even as he wrote to Francis Baily, he knew his health was failing. "Perhaps I may construct a Machine to this extent with my own hands in the course of a few Months if God be pleased to spare my life and I have convenient opportunity." With government sponsorship he could have handed work on a metal machine to an assistant and taken greater care of himself. But the government refused to look at his machine on the grounds that they had spent so much, with no satisfactory result, on Babbage's Difference Engine. So Charles Babbage became the man who, albeit unwittingly, was the cause of Fowler's final disappointment.

The one person who refused to accept defeat, of course, was Fowler himself. Spurred on by excitement over his second machine he continued to work on alone. Having placed his reputation in the hands of strangers, strangers who had ultimately failed him, there was now only one person he could rely on — himself. Throughout the winter and on into the summer of 1842 he devoted himself to creating a full size version of his much improved invention, that long sought after desideratum, a Universal Calculating Machine.

Once embarked on the decimal road, Fowler could not resist a brief diversion into the campaign to nudge England towards a decimal currency. Britain's currency still counted four farthings to the penny, twelve pennies to a shilling and twenty shillings to a pound; a complicated system well overdue for change. But the familiar coins were so well loved that whenever the subject was raised there was such a hue and cry it was immediately dropped.

In 1841, Augustus de Morgan attempted a compromise, suggesting some of the existing coinage could be integrated into a decimal system. He proposed keeping the pound, or sovereign, the half-sovereign and the farthing but discarded the much-loved crown, half crown, shilling and sixpence. A year later Fowler entered the fray with a paper printed for private circulation but which, according to his son Hugh, was seen and approved of by the Governor of the Bank of England. To avoid a reaction Fowler chose to keep the familiar names but alter their values to suit a decimal system. His references to pounds, shillings, pence and farthings exactly corresponded with the values eventually proposed by the select committee, but they renamed them pounds, florins, cents and mil.

Fowler's foray into the decimal currency debate is interesting on two counts. The first is his continuing passion for the kind of mathematical detail that many find a complete nightmare. Attached to his paper was a detailed table of the equivalent values of money in the old and new coinage. Tradesmen needing to adjust the prices of their goods would be lost without such a table. With an entire nation converting, demand would be immense.

The second consideration relates to his new machine. Having bowed to pressure to create a decimal device, his support for a change to a decimal currency was an astute move.

Fowler's hope that he might be spared to construct his second machine was granted, but during the winter of 1842 he began to suffer increasingly from heart problems. He was forced to compromise. Council meetings were held in his offices to allow him to attend, but by the spring of 1843 even this became too demanding. As Caroline and Cecilia handed round the refreshments to his colleagues, he remained upstairs, too ill to take part.

It was a crushing blow. He had responded to all the criticism; the machine was built, better than ever. If only he could demonstrate it again. But the journey downstairs was difficult enough; London was impossible. Someone else would have to promote it. Caroline was intimately acquainted with his work, but choosing her raised complications. In this male dominated society, women were debarred from university degrees or membership of associations such as the Royal Society. It was an uphill struggle for them in any branch of academic life, let alone science — although a few exceptions were made for Ladies with the right credentials.

Augusta Ada Lovelace, daughter of the poet, Lord Byron, was born in 1815. When she was just five weeks old her mother separated from Byron and brought Ada up alone, doing her best to direct her mind away from poetry and towards mathematics and science. Beautiful, charming, aristocratic, a romantic figure and mathematically literate as well, Ada was warmly welcomed by the London scientific community, including Augustus De Morgan. But it is her association with Charles Babbage that is best remembered.

They met at a dinner party where Ada was one of the few people fascinated by Babbage's ideas for his Analytical Engine. Even after her marriage to the Earl of Lovelace, Ada stayed in touch with Babbage. She translated an Italian article on the Analytical Engine, adding her own notes which turned out to be three times as long as the original piece. In these, her grasp of Babbage's work and its potential applications of Babbage's outstripped the minds of many of her male contemporaries.

Not all her activities were quite so esoteric though. Babbage's talent for picking any lock or devising systems for winning at Roulette was well known but Ada developed a more hands-on relationship with gambling. She ran up considerable debts and, like her father, lived life to the full. Sadly, that life was equally short. She died aged thirty-six.

Caroline Fowler also had an enviable grasp of mathematics and calculating machine mechanisms but she lacked one fundamental qualification — a position in society.

The task of promoting Fowler's machine therefore fell to his eldest son, Hugh, then aged twenty-seven. In delegating to Hugh, Fowler overcame a major drawback. Hugh was 'one of them', a privately educated Cambridge scholar and a priest; a background that held the hope of action instead of empty gestures.

Hugh Fowler

There remained one other matter to be resolved. Despite the impression given in his letter to Francis Baily that he was "perfectly free and willing to do anything that was required" to publicise his machine, Fowler had still not made any plans or descriptions available for general scrutiny. Without these, there was a very real danger his second machine would also be side-lined.

This time, he did look to Caroline. As winter turned to spring she sat, pen in hand, ready to commit her father's description of his calculating machine to paper. Somehow, while in great pain, Fowler found the will to dictate a dozen or more pages of complicated description to Caroline. So strongly did he believe that his fame would rest on his calculating machine, albeit posthumously, that he devoted his few remaining days to this task. It is difficult to believe that he would have made this supreme effort if any plans already existed.

In 1835, when Fowler started out on the long road to claim recognition for his machine, Halley's Comet made a dramatic appearance in the skies over Torrington. Eight years later the end of his quest was marked in similar fashion:

> *Woolmers Gazette.* March 25 1843. Comet seen in sky. An Unusual and extraordinary appearance was observed from Barnstaple and neighbourhood on Friday evening last. It consisted of a long and

perfectly defined stream or brush of a reddish-yellow light, in the south-west quarter of the heavens, slightly curved downwards, and extending over a space of about twenty-five degrees, from a little from the west of Sirius through the foot of Orion. It was first seen just after sunset and remained visible until nearly nine, when, as the moon rose, it gradually faded away and disappeared.

Just six days later, on March 31st 1843, the doctor was called to rooms above Fowler's Printing Office to perform his last duty for a respected friend and colleague. At sixty-five years of age, Fowler died of dropsy of the chest, an accumulation of fluid on the lungs associated with heart failure.

There has to be a question mark over the role his work on the calculating machine played in his untimely death. He was driven by the need to answer criticisms of his invention; to prove its significance. That need pushed him into long hours of physical labour in a freezing workshop instead of opting to care for himself. But for the man of conviction that he was, there really was no other choice he could make.

A few days later, family and friends filled St Michael's church for the funeral to celebrate the life of a remarkable man. From humble beginnings, Thomas Fowler had certainly made his mark.

Obituary

On Friday morning, the 31st ult, Mr. T. Fowler, banker and stationer of Great Torrington, after a severe and lingering illness — Mr. Fowler's loss will be deeply felt by his surviving family and by his numerous circle of friends in Torrington and elsewhere — by those especially who could appreciate his truly benevolent and disinterested disposition, and his extraordinary talents, and varied and extensive acquirements in almost every branch of study, but more particularly in mathematics. His aptitude for these studies developed itself at a very early age. He is known, for instance, to have mastered Sandersons Fluxions, entirely without assistance, before he attained the age of 16, any mathematician must know how many previous difficulties must have been surmounted to enable him to do this. There can be no doubt, had he enjoyed the advantage of a regular education, that he would have attained the highest eminence as a man of science; and even as it is, he has, amidst the cares and anxieties of business, accomplished enough to render it unlikely that his name will be easily forgotten.

In the year 1829 he took out a Patent for an apparatus for heating hot houses by water constructed on a principle in hydrostatics which he discovered entirely by his own resources; this principle is now universally adopted. The descriptive pamphlet he published respecting the Thermosiphon has been greatly admired for the clearness and perspicuity with which his views are put forward.

In the year 1838 he published a set of tables for the purpose of facilitating the apportionment of payments among the different parishes of Poor Law Unions (which tables are now in actual use in many Unions.) and in connection with them he soon afterwards commenced what may be considered his crowning effort — the construction of his calculating machine, Notices of this ingenious piece of mechanism have from time to time appeared in the Journal. Papers respecting it have been read before the Royal Society and, in the year 1840, Mr. Fowler himself exhibited it in London before a number of the most eminent scientific men of the age (among them Babbage, Baily, De Morgan etc) by whom, as also by the members of the British Association to whose examination it was submitted at their late meeting at Plymouth, it was, though then in comparatively an imperfect state, highly approved and admired.

Happily Mr. Fowler's life has been spared long enough for the construction of an entirely new machine possessing far greater capabilities than the former one, and it is to be hoped that the result of his anxious and unwearied labours of mind and body — labours which his family have reason to fear contributed to shorten his days — will not long be kept from the world.

Exeter Flying Post 6.4.1843

Thomas Fowler died a profoundly disappointed man, deeply saddened that all his efforts to gain recognition for his calculating machine had failed. His son Hugh was later to say "It is sad to think of the weary days and nights, of the labour of hand and brain, bestowed on this arduous work, the result of which, from adverse circumstances, was loss of money, loss of health, and final disappointment."

Sometime after he was laid to rest in the west side of Torrington graveyard, Mary, returned to the printing office with Cecilia and Caroline to complete one important task. The paper Thomas had dictated to Caroline on his deathbed was still to be printed, a commission that might yet bring the recognition he longed for.

Chapter 11

[+ + −]

Bitter-Sweet Acclaim

Wooden Computer Invented in North Devon. When this improbable headline launched me on the search for Thomas Fowler, I had little idea where it would lead. Hugh Fowler now pointed the way, speaking of this final paper as being "of painful interest to me as having been dictated by my father to my sister on his death-bed, while in great suffering from the disease of which he soon after died." Through the pain of his last days, Fowler's belief in his invention remained unshakeable. The least I could do was to find his dying words and use them to recreate his machine.

But this paper proved one of the most elusive documents to track down. Hugh mentioned King's College so I began with a dedicated trawl of their archives. At first there was nothing, but days of scanning finally paid off. There in black and white was the entry: Wheatstone Collection — 8 reprints of T. Fowler's Calculating Machine. Success! With a full description there was a chance of beginning a reconstruction.

The archivists at King's could not have been more helpful. The requested collection was placed on my desk and I hurriedly thumbed through for the promised reprints. Again. And again. But no matter how often I looked, they were not there. It was devastating to have come so close and have to leave with nothing.

At home in Devon it was back to the drawing board. The pamphlets had been printed. I knew that. Where would the family send them? Two names came to mind. Francis Baily and Augustus De Morgan. But I had been through Baily's correspondence with searing efficiency on the hunt for Fowler's letter. The description was not there. This just left De Morgan.

Much of De Morgan's archive is held at the University of London. Soon I was back in London, hurrying to the Palaeography room at the University library. Amazingly a search on Fowler revealed a reference to a paper entitled *Description of the table part of the new calculating machine invented by Thomas Fowler of Great Torrington, Devon, in 1842*. My hunch had paid off. De Morgan had a copy of the deathbed description.

But it was King's College all over again. The paper was nowhere to be found. Having come so close twice now, this was a crushing blow. I was faced with the prospect of leaving London yet again without the description. Where else might I look? The Royal Society? Unlikely given Fowler's failure there. Babbage's archive? Possibly.

The expertise of archivists in all the libraries I visited has been invaluable during the search for Fowler's story. On this occasion, the archivist at the Palaeography room at University College kept searching — first for the document; then for me. The phone rang as I was about to leave London. He had found the deathbed description.

After years of searching I finally held in my hand the only known description of Fowler's unique calculating machine.[24] I spent the next few hours immersed in a language I barely understood, but it made no difference. Here were Thomas Fowler's own words, at last, reaching across more than 160 years.

A detailed analysis of the paper was for others, but one sentence leapt out. At the bottom of the title page I read "And which is now open for public inspection in the Museum of King's College, London". The machine was to have its moment of glory, and in very influential company.

---

[24] See Appendix 12

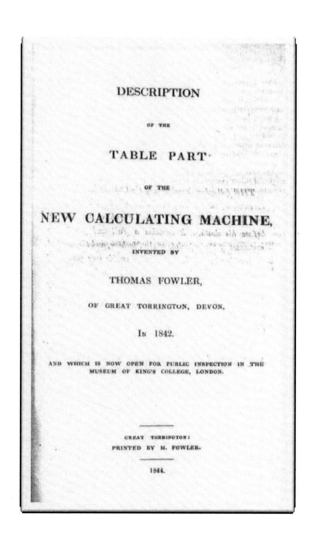

DESCRIPTION

OF THE

TABLE PART·

OF THE

NEW CALCULATING MACHINE,

INVENTED BY

THOMAS FOWLER,

OF GREAT TORRINGTON, DEVON,

IN 1842.

AND WHICH IS NOW OPEN FOR PUBLIC INSPECTION IN THE
MUSEUM OF KING'S COLLEGE, LONDON.

GREAT TORRINGTON:
PRINTED BY H. FOWLER.

1844.

The George III Museum officially came into being on June 22nd 1843, when the Queen presented the King's collection of mechanical models and philosophical instruments to King's College. A special museum visitor's book was signed by Prince Albert at the opening. Six months later this same volume was placed in front of a young man on a visit to the Museum from his home in North Devon. Fowler's nephew, also a Thomas, was visiting the Museum at the invitation of Professor

Moseley, a member of the Engineering Department and acquainted with Fowler following their meeting at Plymouth in 1841.

On December 8th 1843, the Council at King's College decided "that upon the recommendation of the Professors in the Engineering Department, permission be given to deposit Mr. Fowler's calculating machine in the Museum of George III".

On June 18th 1844, Hugh Fowler and his younger brother, Charles, arrived at the museum with the calculating machine to continue their father's work. As with all the previous demonstrations, the assembled group of scientific men were impressed with "the rapidity and accuracy with which [the machine] brought out long sums in multiplication and division as far as ten figures by ten figures". A proud moment for Hugh, but probably clouded by the realisation that exhibited in a glass case in the centre of the Museum was the very machine that was the prime cause of his father's downfall — Babbage's Difference Engine.

When Mr. Goulburn, the Chancellor of the Exchequer, wrote to Babbage informing him that the government was cancelling all support for the Difference Engine, Goulbum offered to return full ownership of the completed portion of the machine to him. Babbage declined and went on the offensive, demanding a public statement that he had no part in the non-completion of the machine. Having suffered personally and financially because of the government's prevarication, he resented any suggestion that he had benefited from the public purse.

While he waited for a decision, the completed portion of the Difference Engine became quite a party piece at Babbage's Saturday evening soirées. However, in 1843, the government decided to place it on exhibition at the George III museum. Hearnshaw's *History of King's College* offers an interesting reflection on how some viewed this remarkable device:

> The museum, it may be remarked, was proving, as museums usually do — a magnet which tended to draw to itself all those pieces of large lumber in London and the provinces which private persons found too good to destroy, inconvenient to keep, impossible to sell, and unsuitable for wedding presents. The biggest white elephant, however, dumped down upon the college during this period was Babbage's calculating machine.

How time and perceptions change. The man now referred to as "the father of computing" was then labelled as the creator of the biggest white elephant ever dumped on the college. The Museum Curator's Diary conveys quite a different impression. Throughout 1850, students and visitors continued to operate the machine. Apart from occasional cleaning and resetting, this finished portion of the Difference Engine was working perfectly seventeen years after Babbage stopped all development on it.

However, the government quietly pursued a policy of ignoring the Difference Engine and its inventor. At the Great Exhibition of 1851, while the Newbury Coat had pride of place, the Difference Engine was nowhere to be seen. The government was, according to Babbage "as insensible to the greatest mechanical as to, what has been regarded as some, the greatest intellectual triumph of their country". Not only this, but the government also refused to release the Difference Engine for exhibition in New York, Dublin and the 1855 French exhibition where *instruments-de-precision* were actively sought.

When the Difference Engine was finally released for display at the International Exhibition of 1862 Babbage expressed his hope that the machine would be prominently displayed alongside another Difference Engine invented by George and Edward Scheutz. He was convinced that, together, they would become one of the greatest attractions in the building. But this was seen as another attempt at self-promotion rather than a simple statement of fact. The Difference Engine was relegated to a small alcove where only three people could see it at a time.

Babbage offered endless drawings and notations to enhance the exhibit, but they required up to 800 feet of wall space, and the decision had already been taken to display an oilcloth. This was the final straw for Babbage. "It is evident that the Royal Commissioners were much better qualified to judge furniture for the feet than of furniture for the head." Having been snubbed, he refused to pay six shillings a day to engage someone to demonstrate the machine and answer questions. In Babbage's opinion the authorities should have recognised the value of the exhibit and taken on this expense. By selling copies of a pamphlet he could have easily covered these costs and more, but a principle was at stake. It seems he preferred to wallow in the perceived injustice of this snub rather than do something about it.

Despite his frustration at both government and public failure to appreciate the full significance of his machine, Babbage continued to work on his Analytical Engine. There are echoes of Fowler's experience in his words written in 1864: "If I survive some few years longer, the Analytical Engine will exist". Sadly by the time he died, just seven years later, only a few fragmentary portions of the mechanism had been completed.

After Hugh and Charles Fowler's demonstration of their father's machine in June 1844, Hugh's career and personal life took over and, with no one actively promoting it, Fowler's machine was forgotten. Seven years later the museum curator was prompted to look and eventually found it in the lumber-room. While Babbage's machine was still in constant use by large numbers of students and visitors, Fowler's was gathering dust in a cupboard. Why this ignominious end to a device he was convinced would be his main claim to fame? There are several candidates for blame.

The Difference Engine was a magnificent piece of precision engineering in step with the modern thinking of the post-industrial revolution. It was almost completely automatic and easily demonstrated by the curator. By contrast, Fowler's machine needed a trained operator to convert the decimal calculation into ternary, enter this in the machine, move the various parts according to a few predetermined rules, read the result and convert it back into decimal. Hugh and Charles could do this but it is unlikely that any museum staff had their skill in operating the machine.

Babbage was a prominent figure in London society, both social and academic. Although work on the Difference Engine ceased over twenty years earlier, he strongly resisted any attempts to remove it from permanent display. By contrast, Thomas Fowler spent his entire life in Great Torrington and relied on the patronage of local nobility to promote his machine in academic circles. After his death in 1843, Hugh appears to have let the momentum go. Given all these factors, it seems almost inevitable that Fowler's machine should be left languishing in the lumber-room.

Hugh made a half-hearted attempt to remove the machine in 1845, but became distracted by events in his own career. Then, in 1849 he was appointed Headmaster at Bideford Grammar School, just eight miles from his childhood home in Torrington. Perhaps being close to family again jogged his memory. In 1851 he wrote to King's College once more.

Sir. You will oblige me by informing me whether the Calculating Machine invented by my late father, Mr. Fowler of Torrington and placed some time since in the Museum of the College, is still there and in good preservation. I ascertained that it was so about two years ago, when I was informed that I might remove it whenever I thought proper, as it had not been presented to the College, but merely placed there for exhibition. I have not yet decided on removing it, as if the College will give it house-room it is as well there for some time longer as in my own premises.

It was this letter that prompted the museum curator to search for Fowler's machine.

Sir. In reply to your letter of the 19th inst. I have called on the curator of our museum to furnish the information you requested, respecting the calculating machine of the late Mr. Fowler. He informs me that it has never been deposited in the museum entrusted to his care on account of its size which would have required more room than could have conveniently been devoted to it. It was consequently placed in a store room of the college and has been protected from all possible injury further than the accumulation of dust. As this store room is intended to form part of another museum for Anatomical preparations it would be desirable whenever convenient to yourself for some arrangement to be made for its removal, as I regret to say that that it is not probable from increasing want of space in the college that room could conveniently be found either for its safe preservation or occasional exhibition.

While Hugh Fowler was ambivalent about whether or not he wanted the machine back, the College, having been reminded of the space it was taking up, was now keen for it to be removed. The machine was taken apart and returned to Hugh's home in Bideford, North Devon. Tragically, it had been broken into so many parts that he found it impossible to reconstruct it again and the pieces were relegated to his attic. Three years later he left the area to take up a new position as Headmaster of the Cathedral School at Gloucester. With his departure went any lingering possibility that Thomas Fowler's hopes for his unique mechanical calculating machine would ever be realised.

Chapter 12
[+ + 0]

The Memory Fades

The final fate of Fowler's calculating machines remains a mystery. The dismembered pieces of the decimal machine were returned to Hugh and were still in his possession in 1875 but the trail goes cold when he died just two years later. Were they passed down through his family? Or perhaps returned to Torrington to be stored together with the remnants of his ternary machine? Their sentimental value was surely far too great for them to be thrown away.

Many of the family remained in Torrington with Mary continuing the business with the same dedication she had brought to it throughout her years with Fowler. Initially she had both Caroline and Cecilia to support her but in 1853, just a year before Hugh's move to Gloucester, Caroline died at the tragically early age of 39. Cecilia gradually took on more and more of the day-to-day work and evidently did very well. She gifted a brass eagle lectern to the parish church and left a substantial sum in her will.

Over the years, the path that led to the now dilapidated barn became less distinct. Here, behind a sturdy planked door held fast with a heavy latch, lay the remains of over forty years of invention. When Fowler's printing shop was finally closed and the family's possessions removed to their new premises in Fore Street — shortly after Hugh's death — did these include the contents of the barn? Was Fowler's first machine carefully removed and stored? Or did it remain in a far corner of the barn loft, dismissed as a worthless pile of timber?

The thought was irresistible. I contacted the Natwest Bank, the then owners of what was Fowler's Printing Shop, for permission to look in the barn. Literally walking in Fowler's footsteps, I followed the overgrown path to the heavy planked door and stepped into the dark interior where Fowler had spent so many long hours in solitude and silence. Over the years I had shared his emotional highs and lows as I read between the lines of his letters. But here I felt closer to him than I had ever been. It was as though these rough stone walls held the memory of Fowler's inspiration. Sadly that was all they did hold. Neither the ground floor nor the loft above held any tangible remnants of his time there.

Perhaps Henry, the only one of Fowler's sons to remain in Torrington, gathered up any remaining pieces. Henry was very much a person out of his father's mould, with a passion for scientific study. After passing his pharmaceutical exams with distinction, Henry set up as postmaster, chemist and druggist at premises in Fore Street, gaining a reputation as an excellent analytical chemist. He was also an amateur archaeologist, closely involved with the opening of an ancient barrow at Huntshaw. He displayed the almost obsessional family trait for detail in his passion for botany by collecting every plant to be found within a radius of ten miles of Torrington. His obituary in the *Transactions of the Devonshire Association* recorded that "There were few, if any, better botanists in the West of England ... Mr. Fowler was endowed with high intellectual capabilities and it was held by all who knew him intimately, that but for the necessity of close application to his multitudinous business pursuits and the care of his large family, he would have been highly distinguished as a man of science".

Like his father before him, Henry also played an active role in the life of the town. When he died, a brass plaque was placed in the parish church "in grateful recognition of the many benefits conferred by him on the inhabitants of his native town by his unselfish energy and perseverance".

Henry's son, another Thomas, stepped into his father's shoes, working alongside his mother as a dispensing chemist and postmaster. In 1888 he was trading as a 'Dealer in Patent Medicine and Perfumery', dispensing such delights as '8 leeches for Mrs. Honeycroft' at a cost of three shillings. He also carried on the family tradition in other ways when he became the church organist, playing the very same instrument his grandfather had worked on so many years before.

These achievements reflected Fowler's determination to provide his sons with the start in life he could only dream of. His eldest, Hugh, won numerous prizes at Blundell's School in Tiverton before gaining a scholarship to Sidney Sussex College at Cambridge. He was ordained then entered education, becoming Headmaster of Helston then Bideford Grammar School in 1850. He obviously had a flair for business as well as education — achieving the impressive combination of raising fees yet still increasing numbers from fourteen to fifty boys. He also boosted the prestige of the school by offering Latin, Greek and other branches of a classical education. Hugh also knew how to motivate pupils, introducing medals for success in examinations made of solid silver and proudly displaying the Bideford coat of arms.

After four years, he took up the post of Headmaster at the Cathedral School in Gloucester, becoming one of the school's most highly regarded masters during the 19th century. Frederick Hannam-Clark, in his book *College School Memories*, writes of Hugh Fowler "There are but few men who possess all the qualities of a first-rate Headmaster. It seems to me, however, that Mr. Fowler held them. He was a happy combination of scholarship, discipline and business … the moderate fortune which he is understood to have acquired during the same period testified to his business aptitude."

Hannam-Clark also paints a vivid picture of the man. "Who that was in the School at that time will forget the wholesome fear he inspired in all the boys, big and little? Who will forget his heavy rapid stride through the school room with his gown flying behind him, or his determinate twanging of his bell on the desk until he got silence? The firm-set mouth, the searching eyes, the dignified bearing, the sonorous voice — all had their effect upon the boys." Characteristics inherited from his father perhaps?

Like his father, Hugh also had eleven children, the eldest of whom, William Weeks Fowler, graduated with a first from Oxford. He became a noted entomologist, headmaster of Lincoln Grammar School, and Canon of Lincoln Cathedral. William also inherited one of his grandfather's passions. His obituary records "In his small and apparently delicate frame, Fowler held a great store of vitality and an apparently inexhaustible appetite for hard work … he had a taste for the not usually attractive labour of collating and tabulating the records of others' results".

Fowler's second son, Charles, graduated from the Royal Academy of Music and made his home in Torquay, South Devon. There he took pupils, composed and gave regular concerts. There are numerous reports of him performing at charity events in North Devon. At one concert in aid of the Infirmary, the *North Devon Journal* reported "The appearance of our talented townsman was the signal for a burst of enthusiasm". He often played his own compositions — on this occasion a Nocturne and Tarantella. Also in his repertoire were *Variations* by Thalberg, which "testing to the utmost the capabilities of the performer, were played in a masterly manner". He frequently played at events to raise funds for the restoration of Torrington's parish church, and composed a chorale for the re-opening in 1862.

Charles returned to London when he accepted a Professorship at the Royal Academy of Music. He became well known for his annual concerts at the Piccadilly home of Miss Burdett Coutts where his patrons included Her Royal Highness, the Duchess of Cambridge. He was justifiably proud of these concerts, publicising them in the *Exeter and Plymouth Gazette*. Lady Clinton and Lady Gertrude Rolle responded, attending one of his concerts at Hanover Square in 1864.

Fowler's fourth son, Newell, followed Hugh with a scholarship to Sidney Sussex College. The local paper reported "The Rev N.V. Fowler MA a native of this town has been elected a fellow of Sidney Sussex College Cambridge on the Blundell Foundation. It were well if the choice of the electors to University honours and emoluments always fell on men who by their scholarship and character as richly merit the distinction as Mr. Fowler." He chose a career in the church, holding a curacy at Towednack in Cornwall before accepting the College living at Ulting in Kent, where he remained until his death.

Twins Paul and Silas were only nine when Fowler died. There is no record of private school or a university education for them, but Paul continued the family tradition by becoming a banker. Silas chose a career at sea, signing up as a secretary's clerk on the *Portland* as soon as he turned seventeen. Fifteen years later he was making headlines. Under the banner "Gallant Rescue", local and national papers alike reported on an incident that befell the steamer *Bombay* in stormy conditions off the south coast of Australia.

There was a tremendous sea running, and two of the crew were washed overboard from the bowsprit. Although the attempt to save them seemed hopeless, a boat was lowered, and Mr. Silas Fowler, the second officer, with six of the crew, volunteered to put off to the rescue. At a distance of more than a mile from the ship, when they were on the point of giving up the search, they providentially came upon the men, and got them into the boat, one of them having his thigh broken. Great anxiety was felt by all in the ship for the safety of the boat's crew, and when they returned on board with the rescued men they were received with loud cheers. A subscription was immediately raised among the passengers and officers of the ship, out of which a sum of money equal to a month's pay was given to each of the men, and a handsome testimonial to Mr. Fowler. On his arrival in England a few days ago, the testimonial was presented to him in the shape of a valuable field-glass with the following inscription: — "Presented to Silas Fowler, Esq., second officer of the Peninsular and Oriental Company's steamer Bombay, by the passengers and officers of the ship, on her passage from Melbourne to Galle, August 1864, for his bravery in saving the lives of two Lascars who were washed overboard.

This was not Silas's only act of bravery. When he was a boy at Bideford Grammar school he saved the lives of two of his schoolmates from drowning in the river and "on many other occasions since he has been at sea, saved or attempted to save life. The Directors of the Peninsular and Oriental Company have promoted him to be chief officer of the *Bombay*; and, a few days ago, they sent an account of the rescue of the two Lascars to the Royal Humane Society, with an application, on Mr. Fowler's behalf for use of the Society's medal. A bronze medal was granted to him by the committee."

In many ways, Fowler's personal legacy, in the intelligence and character cascading through his family, was just as much an achievement as his inventions. All his children, sons and daughters, excelled in their chosen sphere, whether education, music, the church, at sea or in business. And all were highly regarded for their service to others. Fowler would have been proud.

After a lifetime supporting her husband's work, and playing a pivotal part in nurturing their eleven children and holding the business together, Mary died on November 19th 1863. It was her death that led me to the only known representation of Fowler's machine; not buried in the archives in Oxford, Cambridge or London but hidden in plain sight in the one place Fowler knew so well: St Michael's Church in Great Torrington.

His mother's death prompted Hugh to approach the London firm of Heanton, Butler and Bayne with a commission for a commemorative stained glass window, a memorial to all the family members he had lost.

He asked for their names to be inscribed in the glass.

Thomas Fowler, died 1843 Aged 65.
Mary Fowler, died 1863 Aged 72.
Mary Fowler, died 1832 Aged 14.
Thomas Fowler, died 1835 Aged 16.
Caroline Elizabeth Fowler, died 1853 Aged 39.
Frances Honor Fowler died 1859 Aged 32.

But this wasn't all. In discussions over how to make this window a very personal tribute, Hugh requested that two panels be inserted in memory of his father's achievements. They are designed to blend into the overall design, simply colourful panels to the casual observer, but actually a unique memorial to the life and work of Thomas Fowler, mathematician and inventor.

On the left is the Thermosiphon ...

On the right, the calculating machine.

This left me with one question. Twenty-five years after Fowler's death, a stranger was entrusted with the task of creating a visual representation of his calculating machine — a difficult task without drawings. Had a machine survived in a good enough state to act as a model? Or did he work from designs given to him by Hugh Fowler?

Whatever the answer, any original drawings were filed away. Tragically, these were lost when the firm's records were accidentally destroyed seventy-five years later.

Two significant occasions marked Thomas Fowler's work after his death. The first was in 1870 when the family gathered at St Michael's Church to dedicate the memorial window. The second was in July 1875, when The Devonshire Association for the Advancement of Science, Literature and Arts held their fourteenth annual general meeting in Great Torrington. The meeting opened with a concert in the Town Hall, where Charles provided the entertainment. Described as "the eminent pianist of Torquay" by the local paper, he "quite enraptured the audience by his brilliant playing".

But on this occasion it was Hugh who took centre stage. Although their father had died over thirty years earlier, Fowler was still well remembered locally. When the time came for papers to be submitted for this gathering, Hugh grasped the opportunity to celebrate his father's achievements. It was a poignant moment for Hugh when he stood to read, *A Biographical Notice of the late Mr. Thomas Fowler of Torrington, with some account of his Inventions*. From where he stood on the platform in the town hall, he could see directly into the rooms above the printing shop opposite, to where his father's life, so full of promise, ended in bitter disappointment. Hugh began with a brief account of that life:

He was apprenticed at an early age (about 13 or 14, I think) to a fellmonger. It was at this time that his taste for mathematical study began to develop itself ... Few people, if any, of this town or neighbourhood knew, or if they knew, cared, that there was in their midst "a wondrous boy", who, absolutely self-taught, after his hard day's work among sheepskins spent half the night poring over his mathematics ... There was no one, alas! to take him by the hand, and help him to carry on his studies at Cambridge, where alone such talent as he undoubtedly possessed could either have been fully developed or adequately rewarded; for that he would have distinguished himself at the University there can, I think, be no question. So he was left, without help or sympathy, to his solitary studies.

Hugh then used this platform to reinstate his father's claim to his first invention, the Thermosiphon:

> I cannot now speak of the thousand ways in which this discovery, beautiful in its simplicity, has been and is still made available … It only remains that I affirm, without fear of contradiction, that wherever heated fluids circulating through pipes are used for the conveyance of heat, this principle is applied. As I said at the outset, my chief object in coming here today is to prove my father's claim to the honour — a barren honour though it has been to him and to his family — of an invention which is actually now in use in all parts of the world.

Hugh spoke movingly of the great debt of gratitude all Fowler's children owed for ensuring that they received the education he had longed for in vain. Although Fowler never overcame this disadvantage in the eyes of London's academic community, it gave Hugh great pleasure to reflect that his was not a case of a prophet unrecognised in his own land:

> …later in life my father had no cause to complain of want of interest in his pursuits on the part of gentlemen and others of the town and neighbourhood. Some of his kind and sympathising friends are, I think, now present; others have passed away, among whom I may mention the Archdeacon Stephens, Mr. Charles Johnson, Sir Trevor Wheler, and Lord Clinton as having done all in their power to encourage him in his work, and to bring his inventions under public notice.[25]

This they had certainly done, but in the end all their combined efforts failed. Why? Hugh Fowler had his own thoughts. "If the machine could have been constructed in metal, as he so much wished, it might be in use even to this day, the mechanism, unlike that of Babbage, being so simple and yet so effective."

When Babbage died, just four years before this gathering, he had given up all hope of the potential of his Difference and Analytical

---

[25] See Appendix 1

Engines ever being recognised in his lifetime. "If un-warned by my example, any man shall undertake and shall succeed in really constructing an engine ... upon different principles or by simpler means, [a recognition of Fowler's work perhaps?] I have no fear of leaving my reputation in his charge, for he alone will be fully able to appreciate the nature of my efforts and the value of theirs." It was to be over a century before Babbage's reputation was fully restored when the Science Museum in London constructed his Difference Engine No 2 in 1991. From the common ground of their shared nocturnal study of *Ward's Mathematicks*, the lives of Thomas Fowler and Charles Babbage diverged into very different worlds. Yet ultimately background, education and means made no difference — both men died bitterly disappointed at society's failure to recognise and reward their ground-breaking work.

In piecing together Thomas Fowler's story I have come to know him as a man of remarkable intellect, perception and imagination; a man with the rare ability to move beyond accepted reasoning and think creatively, producing beautifully simple solutions to complex problems. Yet his pioneering work was allowed to lapse into obscurity, relegated to a series of index cards in archives across the country. Despite Hugh Fowler's attempts to remind the people of Torrington of their unsung hero, the world yet again failed to grasp the significance of his invention and his memory faded.

Chapter 13
[+++]

Linking Yesterday with Tomorrow

Dr Doron Swade MBE

> In the light of Konrad Zuse's work in Germany on mechanical and electro-mechanical digital devices in the 1930's & 1940's and the almost universal adoption of binary (two state) digital techniques in the electronic computer age, Fowler's calculator was in certain respects vastly more promising than Babbages

With Thomas Fowler's story in place I was left with one overriding question. How could an invention that held so much promise get sidelined and forgotten?

There is no doubt that the Airy's inbuilt prejudice against "these contrivances", was a factor. Without his recommendation for government funding any hope of creating a smaller, metal version of Fowler's machine disappeared. But Fowler was convinced the Royal Society and others dismissed both him and his work because of his humble background; that his chances had been sabotaged before he even set foot in London because he lacked the right connections.

He had a point, but another candidate for blame has to be Fowler himself. His refusal to produce drawings was a frustrating obstacle for anyone interested in his ideas. Visualising the action of the machine from a written description is incredibly difficult, even for the committed reader. A casual enquirer would soon lose interest.

Also, as clever and innovative as Fowler's ternary mechanism was, the need to convert between ternary and decimal was a significant drawback. He tackled this by creating a decimal machine, but his new signed notation still left the operator with a conversion to make at the beginning and end of each calculation.

Another drawback was the use of the carry mechanism as a separate action after the main calculation was complete. In other machines of Fowler's era this process took place as the calculation proceeded. De Morgan highlighted that Fowler's carrying apparatus "which though at present detached, and employed to reduce the result to its simplest form after the main operation has been performed, might without much difficulty be attached to the multiplier or divisor, and work with it." But Fowler never addressed this.

These very practical drawbacks eclipsed the theoretical leap Fowler made and, without an experienced operator to hand, it became inevitable that his machine would end its days in the lumber-room. But Fowler's belief in Ternary as the way forward for mechanical calculation was unshakeable to the very end.

More than a hundred years later another mathematician reached the same conclusion.

Nikolay Brousentsov was born on February 7th, 1925, in the Ukraine. His early studies were disrupted by the First World War, but once victory was declared he set out on a path that took him to Moscow University and Sergei Sobolev. Brousentsov recollected their first meeting. "When I first visited Sergei Sobolev's office, it seemed to me that I was illumined by sunlight — so kind and open was his face. We immediately had a mutual understanding and I am grateful to destiny for leading me to this remarkable man, bright mathematician and knowledgeable scientist, who was one of the first people to understand the significance of computers." Computers. In Fowler's day a computer was the person who sat at a desk performing calculations. There was a long way to go before the name came to refer to the device that has transformed modern life.

After Babbage and Fowler failed to engage the wider world in their passion for mechanical devices, the subject fell into the doldrums.

George and Edward Scheutz from Stockholm had limited success with their own Difference Engine, and de Colmar's Arithmometer continued to achieve modest sales, but it was the mid-1880's before there was any real progress in mechanical calculation. Americans Frank Baldwin and Dorr Felt and the Swede T. Odhner broke the mould with machines that made it into mass production. Then, just before the turn of the century, William Burroughs entered the fray with a machine that marked a turning point in the office calculator industry.

But these were still relatively straightforward hand-powered machines. The next giant stride was taken almost forty years later by such figures as Alan Turing in England, Konrad Zuse in Berlin and John Atanasoff in the USA. Experimental work on electronic calculators culminated in two landmark machines, the Colossus and the ENIAC. Initially invented simply to manipulate numbers, with scientific or military applications in mind, it was thought that only a handful of these electronic brains would ever be needed. Besides, each one was incredibly expensive to build and filled an entire room; hardly a candidate for use in the general office.

A few brilliant minds thought differently, including Sergei Sobolev. In his workshops at Moscow State University, Sobolev quickly identified the potential of this new technology and appointed a team to concentrate on creating a cheaper, smaller, reliable computer suitable for use in the institute laboratories. The man he chose to lead the project was Nikolay Brousentsov.

Brousentsov considered the binary system, identified by American physicist and mathematician John Von Neumann as the optimal choice for the emerging electronic technology. But Brousentsov was not convinced. Daring to be different, he began to explore the possibilities offered by another number system, by ternary. 120 years after Fowler began the very same exploration, Brousentsov also discarded binary in favour of a design based on ternary. It was a seminal moment when, in 1958, his team assembled the first ever ternary computer. Named 'Setun', after a river close to Moscow University, it became the cheapest computer produced at that time in the Soviet Union — perhaps in the world. And cost was not the only advantage. Compared to binary devices, it offered more speed, performed flawlessly and required less equipment and power. Simplicity, economy and elegance — the direct consequences of using ternary.

Despite the fact that these developments took place when the western world was largely ignorant of work being done inside the USSR, export orders were received for 'Setun''. But none made it abroad. The ternary computer was seen as an aberration, non-planned and unacceptable. In the USSR as well across the globe the drive towards binary devices was unstoppable and the 'Setun' computer was sidelined, but not completely forgotten.

In 1970 another Soviet computer specialist, Professor Pospelov, added his voice to those speaking out in favour of ternary. In his mind, the only barriers to introducing ternary systems in computers were purely technical. As soon as economical and effective elements with three stable states could be developed, computers designed to function in the ternary symmetrical number system would follow. [This equates to Fowler's signed, or balanced, ternary number system.]

In the autumn of 1993, Professor Stanislav Klimenko, Director of the Institute of Computing for Physics and Technology in Moscow, travelled to the United States for a lecture tour. His subject was the history of computer science and technology in the Soviet Union. By this time binary had been synonymous with computers for more than thirty years. Yet when it came to questions there was just one subject his audience were interested in, Brousentsov and his ternary computer.

Donald Knuth, Professor Emeritus of the Art of Computer Programming at Stanford University, California, also highlighted the potential significance of ternary. "Perhaps the symmetric properties and simple arithmetic of this number system will prove to be quite important some day when the 'flip-flop' (2 state device) is replaced by a 'flip-flap-flop' (3 state device)" In the April 2002 edition of *The Computer Journal*, Alexey Stakhov shares his view that adopting ternary releases processor capacity resulting in a faster, more reliable machine, particularly when used for 'real time' applications.

Whether these clear benefits will ever overcome the ubiquitous use of binary is a moot point. The stable three-state devices referred to by Donald Knuth and Professor Pospelov are needed to make this a reality. But given these, and the courage to embrace a new approach, ternary machines offering a significant increase in performance over conventional digital computers could become a reality.

The drive for smaller, simpler, cost effective devices continues into the twenty-first century where Thomas Fowler's work may have unexpected relevance. Computer technology is set to make another conceptual leap. In the relentless drive for miniaturisation and ever decreasing costs, scientists are looking to the world of nanotechnology for answers. Current methods limit how small devices can become. But once scientists are able to work at the molecular level, using individual atoms as building blocks, then microscopic computers become feasible. The relevant point here is that these molecular machines need not be electronic. Although not as fast, mechanical molecular devices are potentially much smaller and simpler to design. Significantly, something being explored by Dr Eric Drexler and Dr Ralph Merkle is rod logic, a method that echoes Fowler's choice of rods in his machine.

There have been two contemporary assessments of Fowler's work. In 1993, Dr. Doron Swade MBE, then Senior Curator of Computing at the Science Museum, came across the correspondence between Fowler and Professor Airy and reached the following conclusion:

> With the benefit of hindsight, we can see that Fowler's machine is in certain respects vastly more promising than Babbage's. The use of sliding rods or plates reappeared in the late 1930s and early 1940s when the German pioneer of computing Konrad Zuse built automatic programmable calculators using the technique in his binary (two-state) digital machines. Zuse appears to have had no knowledge of Babbage or of Fowler. In the light of Zuse's work and the near universal adoption of binary for electronic computers, Fowler's machines can be seen to be closer to modern digital circuitry than Babbage's. [26]

'Fowler's calculator is in certain respects vastly more promising that Babbage's'. If only Thomas Fowler were alive to hear those words. The second assessment was contributed in 2002 by Ralph Merkle, Senior Research Fellow of the Institute for Molecular Manufacturing,

---

[26] Swade, D. *The Cogwheel Brain.* London: Little, Brown & Co, 2000

and one of a team of three who made a significant contribution to the development of secure internet transactions:

> Computers might have changed history and our world almost a century sooner than they did had the ideas of Fowler been understood and adopted by Babbage — as they might have been. Babbage's use of a mechanically more complex and unreliable decimal system, instead of the more robust ternary system (almost binary in its simplicity) that Fowler proposed, was a significant factor in complicating Babbage's efforts and, ultimately, in dooming the analytical engine to its fate as just a fascinating footnote in history. On such details of implementation does the fate of new technologies often hang.[27]

For almost two centuries Thomas Fowler's brilliance has lain dormant, inaccessible to those who could have built on his innovative ideas. But perhaps a century here or there is not so critical. With innovators such as Nikolay Brousentsov, Donald Knuth and Ralph Merkle all speaking of the unrealised potential of signed, or balanced, ternary perhaps the time has come to reconsider it as a springboard for the future. The world-wide computer industry may never have the courage to leave binary behind, but neither can they deny that Fowler's choice of ternary was a stroke of brilliance. Perhaps the fundamental significance of his seminal work will yet link yesterday with tomorrow.

> However humble a first idea may be it often leads to the most valuable and refined results, and the very circumstance of giving it honorable publicity is the means of such further improvements as could not a first be anticipated, conclusions such as there may be drawn from the whole history of science, which is only a series of successive improvements on simple and original ideas.

---

[27] Dr Ralph Merkle in correspondence to David Hogan

Chapter 14

[+ − − −]

Reconstruction

How possible is it to picture Thomas Fowler's machines? Without plans it has been extraordinarily difficult to visualise the mechanism or attempt a reconstruction. The small panel in the memorial window in Great Torrington church remains the only known image of Fowler's unique invention. The key word here, of course, is known. Detailed drawings may lie buried in a family archive somewhere, or perhaps a box of curiously figured wooden rods is gathering dust in a forgotten corner of an attic somewhere in Great Torrington, or perhaps Gloucester.

Without the benefit of either of these, the international team that assembled to reconstruct Fowler's machines was presented with the task of working from written descriptions alone.

Detail of the first ternary machine is particularly thin on the ground. From surviving copies of Fowler's Tables and the paper he prepared for the Royal Society, two members of the project team, Dave Hogan and David Batty, made significant progress on Fowler's underlying arithmetic. However, the workings of the machine itself are less clear.

The report in the Transactions of the British Association for the Advancement of Science helps with first impressions. "[The machine] is on the principle of the old abacus, or calculating rods. At the first glance it has somewhat the appearance of a pianoforte, or organ, with all its keys laid bare." De Morgan's paper then paints a more detailed picture describing a machine composed of four similar frames, one of which, the carrying mechanism, was detached. Two of the frames were similar, containing a number of horizontal rods that either influenced or were acted upon by a central, perpendicular frame.

Fowler's correspondence to Airy offer the machine's dimensions of six feet long by one foot deep by three feet wide, and the valuable information that it contained 55 places in the ternary scale. There is more in both documents (available in the appendix) but an attempt to create an actual representation from these was to take a leap in the dark. Fortunately Mark Glusker from California, a collector of historic calculating machines and an interpretive engineer skilled in the realisation of ideas, proved sure-footed. It was a memorable day when on December 9th, 1999 he produced concept drawings of a mechanism that had been lost from public view for over 150 years. Suddenly we were no longer talking of *if* but *when* the machine would be reconstructed.

Mark Glusker's sketches.

The Product, Multiplier and Multiplicand

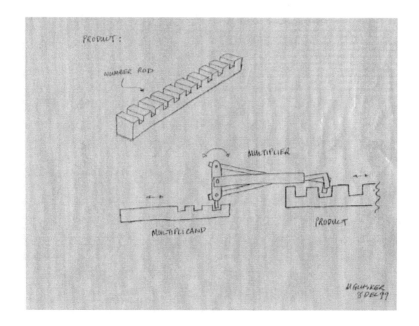

The Multiplier

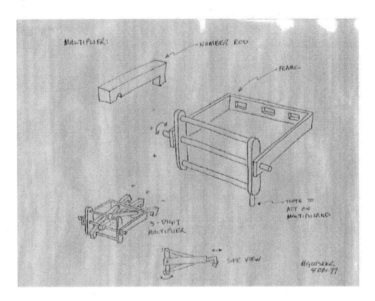

The Multiplicand

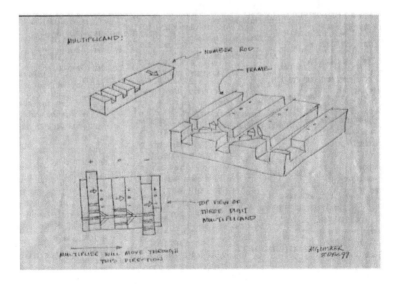

The discovery of Fowler's letter to Francis Baily introduced a fascinating new dimension. Not only were we looking for sufficient information to rebuild Thomas's ternary calculating machine, but also a new and improved version in which the principle of the machine could be "extended into all scales of notation indefinitely, the only limit being the extent of human capability for the mechanical construction." This second machine was described in detail in Fowler's deathbed account and later exhibited at King's College.

Unfortunately at the end of a sequence of detailed descriptions Fowler remarks "By comparing what has been now written with the machine itself, these motions and the construction of the machine will become apparent." If only that were possible! This description also lacked any supporting drawings or descriptions. It seems unlikely that any remain to be found. Would Fowler really have made the supreme effort of dictating lengthy and complicated details of the machine to Caroline when he was so close to death, if he could simply have pointed her in the direction of plans in his workshop?

Early in 2000, the project team gathered in Great Torrington. The town is still easily recognisable as the place Fowler knew. His printing shop has gone but, until very recently, his building (on the far right in the picture below) was occupied by the NatWest, a bank that can trace its history back to Loveband and Co

The Black Horse Inn, the council chamber and, thanks to extensive restoration by the Landmark Trust, John Sloley's house, where Fowler installed his Thermosiphon in the vinery, still appear exactly as Fowler would have known them.

John Sloley's house.

Castle Hill remains a beautiful spot with newly surfaced paths leading down to the Obelisk where Fowler would have stood in 1818 to remember those lost in the Napoleonic Wars. Across the river, Sir Trevor Wheler's home, Cross House, can still be seen standing in parkland on the slopes facing Torrington.

The Pannier Market

Over the past few years this small market town has seen a remarkable community effort. The pannier market has been restored at a cost of several hundred thousand pounds — rather more than the £3,000 Thomas Fowler collected when the market was first built in 1842. A building on Castle Hill has also been restored to provide new facilities for the town. These include a library, tourist information centre, and, for some time, Torrington 1646, where visitors could hear the story of the part the town played in the Civil War. Most significantly, this building originally housed an Information and Communication Technology facility called *The Thomas Fowler Centre*. In this appropriate setting, the project group gathered to witness a seminal moment.

Throughout three years of research, and early attempts at a reconstruction, there was one question in everyone's mind. Were we recreating Fowler's machine, or inventing a modern version of a ternary mechanism? However, when local team member, John McKay, stood to share his understanding of the construction and operation of the machine, all doubts disappeared. Without having seen any of Mark Glusker's concept drawings, John drew an uncannily similar sketch of a critical part of the machine. It was a significant turning point, one that allowed us all to move forward with renewed confidence. Mark, an expert in the use of computer-aided engineering development tools, then applied his unique skills to construct a fully functioning concept model of Fowler's ternary calculating machine.[28]

Mark explains.

It is unusual to attempt a reconstruction of a calculating machine with so little hard data on which to base a design. Claims of historical accuracy become very difficult, if not impossible, to support. Therefore, the model featured on this website[29] was never intended to represent Fowler's first machine exactly as it was built in the 1840s. Instead, it is a concept model of the most simple machine we could envision that satisfies the description written by Augustus De Morgan.

The beauty of the De Morgan description is its brevity — less than a thousand words to describe the machine and its use. Only the essential details are included. When reviewed in conjunction with some basic multiplication problems in balanced ternary, a clear picture of the machine began to emerge. Throughout the reconstruction and over

---

[28] For a detailed account of this process see Glusker, Mark, Hogan, David and Vass, Pamela. "The Ternary Calculating Machine of Thomas Fowler." IEEE Annals of the History of Computing 27:3 (July/September 2005): 4–22. http://doi.ieeecomputersociety.org/10.1109/MAHC.2005.49

[29] http://www.mortati.com/glusker/fowler

numerous transatlantic phone calls, we constantly compared our designs to the description, and eliminated any excess detail that was not mentioned by De Morgan.

The first working model of Fowler's machine in over 150 years was the crude cardboard prototype that I made in late 1999, shown below. Although it was extremely fragile and finicky, it was able to correctly calculate a few simple multiplication problems in balanced ternary.

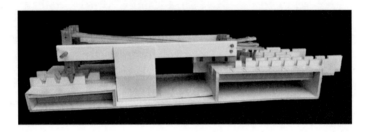

This was a sufficient demonstration to warrant the development of a more precise model..[30]

---

[30] See http://www.mortati.com/glusker/fowler/video.htm for a video demonstration of the operation of the machine.

Our goal is to uncover any additional evidence of how the machine worked, and then use techniques and materials more appropriate to the 19th century to build the entire 55-digit machine that Fowler envisioned.

Other members of the project team have worked on their own visualisations of Fowler's mechanism. One in particular, Torrington man Roy Foster, followed in Thomas Fowler's footsteps in a very practical sense. Inspired by the research findings and Mark Glusker's designs, Roy designed and built his own wooden version of the decimal machine. This photograph of Roy with his model was taken in the council chamber at the town hall. The door that can be seen at the back of the room is one that Thomas Fowler would have walked through on many occasions.

Nothing less than a full reconstruction of both machines is essential if the full significance of Fowler's work is to be realised. Unfortunately the one person who has consistently made this task as difficult as possible has been Thomas Fowler himself. His refusal to produce any plans or diagrams means that much is still left to interpretation.

A concept model was presented by Mark Glusker to Great Torrington in a civic ceremony, prompting a domino run of media articles as his story captured the public's attention. A theatrical presentation of Thomas Fowler's life, written and produced by Pamela Vass, was also given at the Plough Theatre in Great Torrington.

Most significantly, the project team was invited to prepare an article on Fowler for the IEEE Annals of the History of Computing.[31] Pamela Vass and David Hogan were also invited to give a presentation to the Computer Conservation Society at the Science Museum, London. One of Mark Glusker's concept models is now part of the Science Museum's collection and he has also given several presentations on Fowler's work. In 2007, he was invited to speak at the 37th International Symposium on Multiple-Valued Logic, Oslo, Norway. Among lectures at the symposium on such 21st century topics as circuit design, logic functions and quantum computing, the lecture about Thomas Fowler and his nineteenth-century wooden ternary calculator was warmly received. The lecture was accompanied by a demonstration of calculation using the concept model.

Mark gave another lecture and demonstration later the same year at the Arithmeum, a museum of mechanical calculating machinery in Bonn, Germany. Their collection is one of the largest in the world. It consists of over 1200 items, and now includes one of the three original Fowler concept models.

More recently, in October 2012 the Fowler machine was presented again, this time as part of a workshop on calculating machines at Machine Project, an art gallery and performance space in Los Angeles, California. The audience consisted primarily of non-technical people, who were nonetheless fascinated by the idea of a hand crafted calculator made of wood that counted in threes.

---

[31] IEEE Annals of the History of Computing 27:3 (July/September 2005): 4–22.
http://ieeexplore.ieee.org/xpl/articleDetails.jsp?reload=true&arnumber=1498715

A more local invitation has a particular poignancy in the Thomas Fowler story. In 1875, Hugh Fowler stood to deliver a paper on his father to the Devonshire Association for the Advancement of Science. A hundred and twenty-six years later, Pamela Vass was invited to be a guest speaker at the 1999 AGM of the Devonshire Association, held in Great Torrington. Following in Hugh's footsteps, this was an opportunity to again remind the people of Great Torrington of their forgotten genius.

Thomas Fowler's conviction that his original approach signalled the way forward for simple and efficient calculation, an approach vindicated a century after his death, cost him dearly. As his son, Hugh, said "It is sad to think of the weary days and nights, of the labour of hand and brain, bestowed on this arduous work, the result of which, from adverse circumstances, was loss of money, loss of health, and final disappointment. If the machine could have been constructed in metal, as he so much wished, it might be in use even to this day ..."

Fowler's was a tragically untimely death, but for the man of conviction that he was, there really was no other choice but to work on. Nothing could shake his passion for mathematics and his conviction that the beautifully simple principle demonstrated in his machines was worth the price.

I often reflect that had the Ternary instead of the Denary Notation been adopted in the infancy of Society, Machines something like the present would long ere this have been common, as the transition from mental to mechanical calculation would have been so very obvious and simple.

Thomas Fowler May 8th 1841

Illustrations

P 47   Great Torrington Church. Engraved print from study by
        Thomas Allom (1908-1872)

P 48   Organ mechanism By Frederik Magle [CC BY 3.0
        (http://creativecommons.org/licenses/by/3.0)], via
        Wikimedia Commons

P 48   Illustration of Ternary Arithmetic, courtesy of David Hogan

P 49   Babbage's Difference Engine, Science Museum By ©Marcin
        Wichary (www.computerhistory.org) [CC BY 3.0
        (http://creativecommons.org/licenses/by/3.0)], via
        Wikimedia   Commons.

P 51   Charles Babbage. Illustrated London News, 4th November
        1871

P 56   Quote by Dr W. Buckland. By permission of the Science and
        Technology Facilities Council and the Syndics of Cambridge
        University Library.

P 67   Augustus De Morgan. Severino666-commonswikki (talk1
        contribs)

P 75   George Bidell Airy. Frontispiece to autobiography of George
        Bidell Airy. 1896.

P 86   Interior of Plymouth Athenaeum. Thomas Allom (1804-1872)

P 96   Ada Lovelace by Alfred Edward Chanlon - Science & Society
        Picture Library.

P 97   Hugh Fowler. In, *Memories of the College School, Gloucester*,
        Frederick Hannam-Clark 1890. Courtesy of Gloucestershire
        Archives. Ref: B243/7447GS

P 100   Thomas Fowler's gravestone. Photo Pamela Vass.

P 103   Frontispiece to Fowler's deathbed description. Courtesy of
        University College, London

P 110   Barn behind Fowler's print shop, Courtesy of NATWEST.

P 111   Fowler's daughter-in-law and grandson outside Torrington
        Post Office. Fore St, Great Torrington. Courtesy of Devon
        Archives

P 115   Fowler Memorial window. St Michael's Church Great
        Torrington Stained Glass Photograph - Roy Foster

P 116   Panels from the Memorial Window. Photographs - Roy Foster

P 120   Quote from Dr Doron Swade MBE, The Cogwheel Brain.

P128/129 Concept drawings by Mark Glusker

P 130   Torrington Square 2016. Photo Pamela Vass

Appendices & Sources

Appendix 1   Fowler. H., 1875. Biographical Notice of the late Mr Thomas Fowler of Torrington.

Appendix 2   Introduction: Tables for Facilitating Arithmetical Calculations &c. T. Fowler

Appendix 3   Mark Glusker on Fowler's Tables

Appendix 4   Fowler's Binary and Ternary arithmetic David Hogan

Appendix 5   Description of a Calculating Machine invented by Mr Thomas Fowler of Torrington, Devonshire. By Augustus De Morgan, Esq.

Appendix 6   Thomas Fowler on his Calculating Machine. General Notation. Notation for the Calculating Machine. Tables

Appendix 7   Notice of Mr Fowler's New Calculating Machine. Communicated by Professor Airy. Transactions of the BAAS 1840 and Athenaeum Report

Appendix 8   Sir Trevor Wheler/Professor Airy Correspondence

Appendix 9   Thomas Fowler/Professor Airy correspondence

Appendix 10   On a New Calculating Machine. By Mr Fowler. Transactions of the BAAS 1841 and Athenaeum Reports

Appendix 11   Thomas Fowler/Francis Baily Correspondence

Appendix 12   Description of the Table Part of the New Calculating Machine invented by Thomas Fowler of Great Torrington, Devon, in 1842

Appendix 1

Fowler, H., 1875 Biographical Notice of the late Mr. Thomas Fowler of Torrington, with some account of his inventions.

Ref: Trans. Devon Ass. Advmt. Sci. 7. 171-8inc

I have been requested by some kind and indulgent friends here to put together a few memoranda respecting my late father, Mr. Thomas Fowler, of this town, and to throw them into the form of a paper to be read before the Association. After some consideration, and after enquiry whether such a paper would be in accordance with the declared objects of the Association, I have consented to do so. I am informed by our Secretary that papers are expected to contain something new, and that they should not take the form of a popular lecture. I find, however, that biographical notices have sometimes found a place in the proceedings of the Association; and these, as being records of the past, can hardly be expected to contain much of novelty. The main difficulty in dealing with biographies is the avoidance of an excess of dullness. If, therefore, I transgress the first of our Secretary's injunctions, I shall probably not violate the second. After all, I should not have ventured to appear before you if I were not in a position to claim for my father the credit of an invention which is still in universal use. He was indeed, as many present know, a man of high attainments, especially in mathematics and natural science; but distinguished mathematicians and students of natural science are not so rare amongst us that they should in all cases deserve a public notice like this. Mr. Fowler's special claim to notice, I repeat, is that, at least in one particular, he fulfilled Bacon's condition, and produced "fruit" as the result of his investigations. The invention to which I refer is the well-known method of heating buildings by hot water circulating through pipes. The fact that he was the sole inventor and patentee some fifty years ago is not equally well known. You will, I trust, bear with a son's natural pride in his father's genius while I occupy the short time allotted to me, first in giving a few particulars of his life, with some other instances of his inventive power; and then in substantiating his claim to this particular invention, with a short description of its principle.

He was born in this town nearly a hundred years ago, in the year 1777, of humble parentage, his father being a cooper. He received the barest rudiments of education — not more, certainly, that the three R's – at a small school here. He was apprenticed at an early age (about 13 or 14, I think) to a fellmonger. It was at this time that his taste for mathematical study began to develop itself. I have here, and I shall always retain as an heirloom, the very book, *Ward's Mathematician's Guide,* the only one on the subject which he for a long time possessed. This book, as is usually the case with the *homo unius libri,* he thoroughly mastered, and that without the slightest help from any one. No one could have been more entirely self-taught than he was. Mathematicians in those days were very scarce in this part of Devonshire, and probably elsewhere, even in the great centres of education. The country was then lying under the incubus of the French war, and neither this nor any other of the arts of peace could possibly flourish as they have done since. Few people, if any, of this town or neighbourhood knew, or if they knew, cared, that there was in their midst "a wondrous boy", who, absolutely self-taught, after his hard day's work among sheepskins spent half the night poring over his mathematics, until he had gone so far as to master Saunderson's Fluxions, the name by which the method of the differential calculus, as far as it was then known, was designated. There was no one, alas! to take him by the hand, and help him to carry on his studies at Cambridge, where alone such talent as he undoubtedly possessed could either have been fully developed or adequately rewarded; for that he would have distinguished himself at the University there can, I think, be no question. So he was left, without help or sympathy, to his solitary studies. Yet he did not relax on this account; in fact he never wholly laid them aside to the hour of his death. His whole life was spent in Torrington. He established himself here as a printer and bookseller (his printing-machine, by the way, he made with his own hands on a plan of his own invention); and he afterwards became clerk, and then partner and sole manager, of the only bank in the town. I must not now dwell on the deep debt of gratitude which his children, and not one of them more than myself, owe to his memory for the sacrifices he made that they might have the advantages of education which he had himself longed for in vain. It gives me much pleasure to say that, later in life my father had no cause to complain of want of interest in his pursuits on the part of gentlemen and others of the town and neighbourhood. Some of his kind and sympathising friends are, I think, now present; others have passed away, among whom I may mention the Archdeacon Stevens, Mr. Charles Johnson, Sir Trevor Wheler, and Lord Clinton as having done all in their power to encourage him in his work, and to bring his inventions under public notice.

Mr. Fowler's belief, I know, was that his fame would mainly rest on his calculating machine. He first constructed one in the year 1840, and afterwards another, greatly improved, in the year 1842. Being of wood, it was necessarily of large size, filling, as nearly as I can remember, a cubic space of about five feet high by four broad, and the same depth. This machine was for some time exhibited in the museum of King's College, London. I remember myself working it there in the presence of several scientific men, who expressed satisfaction with the rapidity and accuracy with which I brought out long sums in multiplication and division as far as ten figures by ten figures. My father was strongly advised to construct it in metal; and he would certainly have done so if he had had the means. But the cost would have been very great; and this made it impossible. The machine was in London at the time of my father's death. Some time after I was requested to remove it. It was taken to pieces, packed in a case, and sent down to me. I have the *disjecta membra* still in my possession, but in so fragmentary a condition that they cannot again be put together. I have also a printed description, very minutely drawn up, of the mode of construction and working. One of these papers is still of painful interest to me as having been dictated by my father to my sister on his death-bed, while in great suffering from the disease of which he soon after died. The following extracts from a letter addressed to Mr. Francis Baily, a distinguished savant of his day - I believe a F.R.S. - contain my father's account of the capabilities of the machine: "I find it practically useful for all, even the most extensive calculations, for the public funds, interest, commission, ready-reckoning, to the millionth part of a farthing … and even for the calculation of logarithms, or for finding the natural number to any logarithm, which the machine exhibited in London is capable of, in a singularly beautiful and concise manner, to twelve or thirteen places." He speaks also of its adaptability to any scale of notation. It is sad to think of the weary days and nights, of the labour of hand and brain, bestowed on this arduous work, the result of which, from adverse circumstances, was loss of money, loss of health, and final disappointment. If the machine could have been constructed in metal, as he so much wished, it might be in use even to this day the mechanism, unlike that of Babbage, being so simple and yet so effective.* (*Footnote: The government of the day refused even to look at my father's machine on the express ground that they had spent such large sums, with no satisfactory result, on Babbage's "calculating engine" as he termed it.)

Yet calculating machines are no mere philosophical toys. The arithmometer of M.Thomas de Colmar, a description of which I have in this pamphlet ... (Instructions for the use of the Arithmometer, or Calculating Machine, invented by M.Thomas (de Colmar). Paris: L Guerin, Rue du Petit-Carreau. 1863) ... is used extensively in connections with electrical computations; and I have a letter from my friend, Professor Adams, of Cambridge, in which he informs me that one of Thomas' machines is in constant use there in the observatory, and that it is most valuable in shortening the tedious processes of astronomical calculations.

In the year 1838 Mr. Fowler published his *Tables for facilitating Arithmetical calculations, intended for calculating the proportionate charges on the parishes in Poor Law Unions; and useful also for various other purposes.* He was at the time treasurer of the Torrington Union. In the preface he speaks of the trouble of making these calculations by the ordinary method; of his trying common logarithms, and so abridging the labour, but not to any great extent. He then says: "Happily I hit on the idea that any number might be produced by a combination of the powers of 2 or 3, and consequently that the same indices of the powers that produced any two or more numbers would also represent any other two or more quantities bearing the ratios of these numbers one to another respectively. I now saw that my object was attained, and that I had only to form a table of consecutive numbers, with the corresponding indices of the powers of 2 or 3 that would produce them." I do not know whether this method is still in use in Poor Law Unions, as it was for some time in Torrington and elsewhere. There is no doubt, however, that my father's method does greatly facilitate these troublesome calculations.

As it is probably new to many present, I will try to explain it, taking only the binary table, in which we have indices of the powers of 2.

Now any number whose index is 0 will be 1, as in the algebraic formula $x^0 = 1$. Then, since $2^1 = 2$, we will have 2, index 1; 3 will be represented by the two indices 1, 0; 4 is 2^2; therefore we have 4, index 2. As the numbers go on, the indices are written in succession with the greatest ease and rapidity. An easy example will sufficiently illustrate the use of the tables. The example given in Mr. Fowler's book is purposely as difficult a one as could well be supposed likely to occur in ordinary practice, as it goes to farthings and decimals of a farthing.

No.	Indices of the powers of 2.		No.	Indices of the powers of 2.
1	0		32	5
2	1,		33	5, 0.
3	1, 0.		34	5, 1.
4	2		35	5, 1, 0
5	2, 0		36	5, 2.
6	2, 1.		37	5, 2, 0
7	2, 1, 0.		38	5, 2, 1.
8	3		39	5, 2, 1, 0
9	3, 0		40	5, 3.
10	3, 1.			

Let us suppose an average annual assessment on five parishes to be £1,000. Parish A is assessed in £360, B in £230, C in £180, D in £160, E in £70. The amount (say) of £150 is required to be raised among these parishes. Of course a simple problem like this might be more expeditiously worked by the common arithmetical rule of proportionate parts. I only take it by way of illustration.

Indices.	£	s.	d.	
9 =	76	16	0	
8 =	38	8	0	
7 =	19	4	0	
6 =	9	12	0	
5 =	4	16	0	[The power taken for the proposed ratio.]
4 =	2	8	0	
3 =	1	4	0	
2 =	0	12	0	
1 =	0	6	0	
0 =	0	3	0.	

Take some convenient power of 2 (say the fifth) as the third term of the ratio. As £1,000: £150 :: £32: £4 4/5ths, whence we gather that the value of the fifth power in this ratio is £4 16s. A table is then constructed where the terms downwards from the fifth power are successively found by *dividing* £4 16s by 2, and upwards by *multiplying* by 2, as far as may be needful. Now the assessment of parish A being £360, we look out this number in the tables, and find the indices to be 8, 6, 5, 3. Taking their values from the table, we have:

Index.	£	s.	d.
8 =	38	8	0
6 =	9	12	0
5 =	4	16	0
3 =	1	4	0

£54 0 0 The contribution of parish A.

So, B's assessment being £230, the indices of which number are 7, 6, 5, 2, 1, we have as B's contribution £34 10s. In the same way, C's assessment being £180 (indices 7, 5, 4, 2), its payment will be £27; that of D (£160, indices 7, 5), £24; that of E (£70, indices 6, 2, 1), £10.10s. Then £54+£34 10s+£27+£10 10s = £150 as required. The tables are in fact a simple and ingenious modification of the method of logarithms.

I now come to the thermosiphon. Its principle is very simple, and I shall not occupy your time long in describing it. The diagram before you represents it in its most simple form, but not as it is applied to any particular purpose.

A and B are two open vessels, of which A is set over a fireplace. They are united by the connecting-tube D. CC is a tube bent into the form of a siphon, and so suspended that its ends are immersed about half-way in the water in the two vessels. There is a cock at G, through which the siphon is filled. The end in the vessel A is bent with its orifice upwards, to prevent the air-bubbles from the bottom of A, when heat is applied, from going into the tube and lodging in the upper part. I need not dwell on the details of the mode of setting the apparatus at work, stopping and opening cocks, and so on; I will go at once to the *rationale* of the process. You know that in the common siphon with two unequal legs the flow of water out of the vessel in which the shorter leg is immersed is due to the difference of the weights of the columns of water in the two vessels acting against the atmospheric pressure. In the thermosiphon the legs are of equal length. Here therefore, the pressure of the atmosphere being equal at both surfaces, there is perfect equilibrium as long as the temperature in the two vessels is equal. Now if heat be applied to vessel A, the water will expand. The fluid at the end of the tube in A becomes specifically lighter than that in the other end in vessel B; this at once destroys the equilibrium.

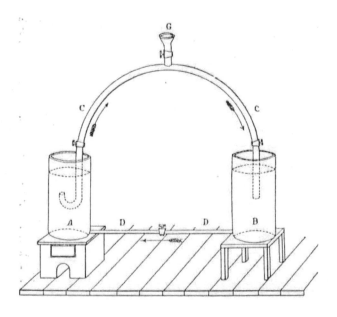

The tube begins to act as a siphon; the warm fluid rises, and the cold water descends in B, and flows through the connecting tube D into A. The circulation then goes on with increasing rapidity, the more if the descending leg be kept as cold as possible. The extreme height to which the fluid will rise in the thermosiphon is of course limited by the force of atmospheric pressure to about thirty feet.

I cannot now speak of the thousand ways in which this discovery, beautiful in its simplicity, has been and is still made available. I may, however, mention one way in which, interesting though it is by way of illustration, the thermosiphon has *not*, I venture to say, been often used. Mr. Petherick, of this town, who was employed by my father at the time to fix the apparatus at Castle Hill, Bicton, and elsewhere, tells me that he remembers a leg of mutton being boiled in a vessel placed forty feet from the fireplace, oil being used as the circulating fluid, and the pipe containing it made to pass through the water in the vessel in which the mutton was boiled. I need not remind you that boiling oil is hotter than boiling water.

It only remains that I affirm, without fear of contradiction, that wherever heated fluids, circulating through pipes are used for the conveyance of heat, this principle is applied. As I said at the outset, my chief object in coming here today is to prove my father's claim to the honour — a barren honour though it has been to him and to his family — of an invention which is actually now in use in all parts of the world. I take my stand on the date of the patent which I here exhibit to you, the year 1828; and I affirm that the mode of heating buildings with hot fluids, as described in the specification appended to the patent, and in this pamphlet,* was at that date absolutely new and unknown. My father met with the proverbial fate of inventors under the then wretched state of the Patent Laws. The invention was very soon pirated in all directions. There was no remedy but by costly legal proceedings; and even if he had had the means to conduct them, success would have been very doubtful; for the slightest modification of the principle of an invention seems then (I do not know how the law stands now) to have been sufficient to bar the penalties of infringement. But surely it is a monstrous thing that unscrupulous men should be able to make large fortunes, as they so often have done with impunity, by picking other people's brains. *Sic vos non vobis!*

A Description of the Patent Thermosiphon, with some modes of applying it to horticultural and other useful and important purposes. London: Longmans. 1829

Appendix 2

TABLES FOR FACILITATING ARITHMETICAL CALCULATIONS

By Thomas Fowler Treasurer to the Torrington Union.

TABLES

FOR FACILITATING

ARITHMETICAL CALCULATIONS,

INTENDED

FOR CALCULATING THE PROPORTIONATE CHARGES
ON THE PARISHES

IN POOR LAW UNIONS,

AND WHICH ARE ALSO

USEFUL FOR VARIOUS OTHER PURPOSES.

DEDICATED BY PERMISSION TO
THE
RIGHT HONOURABLE LORD CLINTON,
CHAIRMAN OF THE BOARD OF GUARDIANS OF THE
TORRINGTON UNION.

BY THOMAS FOWLER,
TREASURER TO THE TORRINGTON UNION.

LONDON:
PRINTED FOR LONGMAN, ORME, BROWN, GREEN, AND LONGMANS,
PATERNOSTER-ROW.
1838.

INTRODUCTION

The following TABLES are Published chiefly for the purpose of facilitating the very troublesome Calculations, which occur every Quarter in making up the Accounts of POOR LAW UNIONS. Having myself been employed in the TORRINGTON UNION, to make up the Accounts, at the commencement, I found those Calculations the most troublesome part of the business, and had recourse to the common Logarithms, which certainly abridged the labour, yet even with this valuable aid I was not satisfied, and was constantly searching after some other method more simple, and of easier application: — happily, I hit on the Idea, that any Number might be produced by a combination of the Powers of the Numbers 2 or 3, and consequently, that the same Indices of the Powers that produced any two or more Numbers, would also represent any other two or more quantities bearing the Ratios of these Numbers, one to another respectively; – I now saw that my object was attained, and that I had only to form a Table of consecutive Numbers, with the corresponding Indices of the Powers of 2 or 3 that would produce them, this was quickly done, by taking the Binary and Ternary Scales of Notation, as the foundation of my proceedings, and I was much delighted at observing the beautiful order in which the Indices introduced themselves into the Tables, and which made their construction a matter of great facility.

Since the Tables have been constructed, the Calculations for the Torrington Union have been made by one or other of them, and the Calculator finding his Work so easy, and so materially abridged, soon caused the Tables to be noticed by the Board of Guardians, of which the Nobleman, who has so kindly permitted his Name to appear on the title Page, is Chairman; – His LORDSHIP, and the very efficient and respectable Auditor of the Union Accounts, C. W. JOHNSON ESQ., so much pleased with this abridgement of the Calculator's Labours, that they immediately desired me to publish them.

I believe the Plan is entirely new, although the properties of the Powers of the Numbers 2 and 3, of which I have now availed myself, have always been known to Mathematicians, and doubtless, the want of some particular object for their use, such as that which is now furnished by the really beautiful mode of keeping the Accounts of the Poor Law Unions, is the reason that an application so obvious and simple, has not hitherto found its way in to the various treatises on Arithmetic.

The necessary Forms of application of the Tables to some useful purposes, will be found in the proper place, and I venture to hope that an intelligent Public will favourably receive this little Work; it is published in the hope of being useful to a valuable Class of Men, by one who is well acquainted with long and laborious Calculations.

In the course of my observations on the Binary and Ternary Scales, I have fallen on a species of Binary and Ternary Arithmetic, which appears to possess some curious Properties, but, as writing on this subject is incompatible with my present purpose, I have only given a short Example of Multiplication, in what may be termed, Ternary Arithmetic. The process is extremely easy, and may be extended to very large Numbers.

Should the Sale of the present Edition be favourable, another will soon follow, in which the Ternary Table will appear under another Form, and extend from Unity to Numbers almost indefinitely great, and also contain some other curious and I hope, useful matter.

THOMAS FOWLER

Great Torrington, Devon,
November 15th 1838

Fowler's Tables continue with sections headed:

- Description and Use of the Binary and Ternary Tables
- Explanation of the signs
  - o   Of the Binary Table
  - o   Of the Ternary Table
- Examples
- Observations
- Application of the Tables, Etc.
- Example of multiplication
- Binary Table – Or a Table of Indices of the Powers of the Number 2 that produce all numbers from 1, to 130048
- Ternary Table – or , a Table of Indices of the Powers of the Number 3 that produce all numbers from 1, to 3985807

Appendix 3

Mark Glusker on Fowler's Tables

A sample page of the ternary tables is shown below.

40

No.	Indices of the Powers of 3.		Leading Numbers and their Limits.
	+	−	
620	6,	4,3,0,	451616 $^-$
621	6,	4,3,	⊕ 452709
622	6,0,	4,3,	453802 $^+$
623	6,1,	4,3,0,	453803 $^-$
624	6,1,	4,3,	⊕ 454896
625	6,1,0,	4,3,	455989 $^+$
626	6,2,	4,3,1,0,	455990 $^-$
627	6,2,	4,3,1,	⊕ 457083
628	6,2,0,	4,3,1,	458176 $^+$
629	6,2,	4,3,0,	458177 $^-$
630	6,2,	4,3,	⊕ 459270
631	6,2,0,	4,3,	460363 $^+$
632	6,2,1,	4,3,0,	460364 $^-$
633	6,2,1,	4,3,	⊕ 461457
634	6,2,1,0,	4,3,	462550 $^+$
635	6,	4,2,1,0,	462551 $^-$
636	6,	4,2,1,	⊕ 463644
637	6,0,	4,2,1,	464737 $^+$
638	6,	4,2,0,	464738 $^-$
639	6,	4,2,	⊕ 465831
640	6,0,	4,2,	466924 $^+$
641	6,1,	4,2,0,	466925 $^-$
642	6,1,	4,2,	⊕ 468018
643	6,1,0,	4,2,	469111 $^+$
644	6,	4,1,0,	469112 $^-$
645	6,	4,1,	⊕ 470205
646	6,0,	4,1,	471298 $^+$
647	6,	4,0,	471299 $^-$
648	6,	4,	⊕ 472392
649	6,0,	4,	473485 $^+$

158

For each decimal number in the leftmost column, the next two columns identify the powers of three, both positive and negative, to express that number in balanced ternary. The column at the far right is used to expand the range of the table by a factor of 3^6. Each number in the centre of the block of three can be expressed in balanced ternary by adding six to the value of the indices shown to the left. The number directly above and below indicate the range over which this value can be used. For example, numbers between 1094 and 3280 can be determined by converting the middle number, 2187, into balanced ternary, and then looking up the remainder in the left-most columns of the table, keeping in mind that the remainder may actually be a negative value.

An Example of Calculating With Fowler's Ternary Tables

In Fowler's book, Tables for Facilitating Arithmetical Calculations, the section "Application of the tables &c" gives an example of his use of the ternary tables to simplify the Poor Law Union calculations. Here Fowler states that the assessment of the whole Union is £7416, and the establishment charges for the whole are £177/2s/1.5d. He also states the individual assessments for several of the parishes. I will use Fowler's example throughout this explanation.

The problem Fowler needs to solve is: if the individual parish assessments are known, and the assessment for the whole Union is known, given the total establishment charge for the entire Union, how much should each individual parish pay?
Let A = the individual parish assessment (value of the parish)
Let B = the assessment of the whole Poor Law Union (value of the entire Union)
Let C = the establishment charge the individual parish must pay (fee for the parish)
Let D = the total establishment charge for the whole Union (total fees for the entire Union)

Each parish should pay proportionally, according to how large its assessment is. The quantity "A/B" is a measure of how much an individual parish is worth as compared to the whole. This fraction should equal the amount the individual parish pays as compared to the total charges, as measured by the quantity "C/D". In other words, if a parish is worth one-quarter of the value of the whole Union, then it should pay one-quarter of the total establishment charges for the Union.

In mathematical terms, $A/B = C/D$

To simplify the equation and solve for C, multiply both sides of the equation by "D": $C = D \times (A/B)$

Due to the difficulties of calculating in the old British monetary system, all monetary values were converted to farthings before any other calculations were done. After the calculations were complete, the values were converted back to £sd.

Here are the steps to perform this calculation using the decimal system:

- ☐ convert each value of "A" to farthings (calculate once for each parish)
- ☐ convert "B" to farthings (single value — only performed once)
- ☐ convert "D" to farthings (single value — only performed once)
- ☐ calculate "A/B" (calculate once for each parish)
- ☐ multiply "A/B" times "D" to obtain the value for "C" (calculate once for each parish)
- ☐ convert "C" from farthings to £sd. (calculate once for each parish)

There is a lot of multiplication and division involved, and since there are 960 farthings per pound, calculating in farthings means the numbers are usually very large.

Fowler probably had to perform these calculations repeatedly, so any reduction in the number of steps, or simplification of the individual steps, would be a great help. This is where his ternary tables come into play.

Fowler describes making a chart of the charges that would apply to parishes if they were assessed at exact powers of three (i.e., £1, £3, £9, £27, £81, £243, £729, £2187 etc). He calls this chart the "New Series of Corresponding Value With the Like Terms of the Ternary Scale". To build this chart, Fowler begins with a calculation for a hypothetical parish, which has been assessed at value equalling an exact power of three, in this case, the 6th power of three, or £729 (this should be some suitably large power of three — I believe the choice of which power to use is somewhat arbitrary, larger powers require more calculation to make the initial table, smaller powers may not have enough resolution at the lower end).

Let A = the hypothetical parish assessment, (£729)
Let B = the assessment of the whole Union, (£7416)
Let C = the amount of establishment charge for a single hypothetical parish (unknown)
Let D = the total establishment charge for the whole Union (£177/2s/15d or 170022 farthings)

We are still calculating in decimal, so Fowler uses the equation described above to calculate the value of "C" for the hypothetical parish: $C = D \times (A/B)$

The resulting value of C for an assessment of £729 is 16713.33 farthings, or £17/8s/2.25d with a remainder of 33/100ths of a farthing. This number forms the entry in the table for the sixth power of three. The other values in the chart are formed by successively multiplying or dividing this value by three. The chart is shown on page xvi, and you can confirm that £17/8s/2.25d is three times £5/16s/0.75d which is three times £1/18s/8.25d, and so on.

Given this table, how does one calculate the establishment charges of a parish if its assessment is not an exact power of three?

First, one must convert the assessment value to signed ternary, a task made easier by the fact that the assessments are rounded to the nearest pound. For the Great Torrington parish, assessed at £1207, Fowler's ternary table shows positive indices for 7 and 0, and negative indices for 6, 5 and 2. This is simply a different way of expressing the signed ternary number "+--00-0+". Another way of expressing the same thing: 1207 = 2187 – 729 – 243 – 9 + 1

It then follows that the establishment charges for a parish assessed at £1207 would be as follows:
PLUS the charge for a parish assessed at £2187 (positive index 7)
MINUS the charge for a parish assessed at £729 (negative index 6)
MINUS the charge for a parish assessed at £243 (negative index 5)

MINUS the charge for a parish assessed at £9 (negative index 2) PLUS the charge for a parish assessed at £1 (positive index 0) the total will equal the charge for a parish assessed at £1207

The values of the charges for each of the powers of three have already been calculated in the chart on page xvi, shown below.

£.		£.	s.	d.		£.	
7416	:	177	2	1½	::	729	: ?

```
        20
      ─────
       3542
         12
      ─────
      42505
          4
      ─────
     170022
        729  = 3⁶
      ─────
    1530198
     340044
    1190154
  ──────────── Farthings.
7416 ) 123946038 ( 16713,33 =
       7416
      ─────
      49786
      44496
      ─────
      52900
      51912
      ─────
       9883
       7416
      ─────
      24678
      22248
      ─────
      24300
      22248
      ─────
      20520
      22248
      ─────
      ─1728
      ═════
```

TABLE.

OR

NEW SERIES,

OF

CORRESPONDING VALUE

with the like Terms

OF THE

TERNARY SCALE.

THE RATIO,

£7416 : £177 : 2 : 1¼.

Indices	£.	s.	d.	qrs.
8,=	156	13	8	¾,97
7,=	52	4	6	¾,99
6,=	17	8	2	¼,33
5,=	5	16	0	¾,11
4,=	1	18	8	¼,04
3,=	0	12	10	¾,01
2,=	0	4	3	½,34
1,=	0	1	5	0,78
0,=	0	0	5	½,92

In the above Table, the Term represented by Index 6, is found by the present operation; every other Term downwards is found by successively dividing by 3, down to Index 0, and by multiplying upwards by 3, as far as needful; beginning in both cases at the Term £17 : 8 : 2¼,33, represented by Index 6.

On the top of page xviii, shown below, the example for Great
Torrington shows the calculation of the establishment charges for that
parish.

xviii.

1207 **GREAT TORRINGTON.**

	Ind.	£.	s.	d. qrs.		£.	s.	d. qrs.
2187	=7=	52	4	$6\frac{3}{4}$,99	Index, 6=	17	8	$2\frac{1}{4}$,33
	0=	0	0	$5\frac{1}{2}$,92	5=	5	16	$0\frac{3}{4}$,11
−980					2=	0	4	$3\frac{1}{2}$,34
		52	5	$0\frac{1}{2}$,91				
	−23	8	$6\frac{1}{2}$,78		−23	8	$6\frac{1}{2}$,78
		28	16	6 ,13	for Great Torrington.			

265 **LITTLE TORRINGTON.**

Index, 5=	5	16	$0\frac{3}{4}$,11	Index, 2= −0	4	$3\frac{1}{2}$,34
3=	0	12	$10\frac{3}{4}$,01			
1=	0	1	5 ,78			
0=	0	0	$5\frac{1}{2}$,92			
	6	10	$10\frac{1}{4}$,82			
−0	4	$3\frac{1}{2}$,34				
	6	6	$6\frac{3}{4}$,48	for Little Torrington.		

120 **ALVERDISCOTT.**

Index, 4=	1	18	$8\frac{1}{4}$,04
3=	0	12	$10\frac{3}{4}$,01
2=	0	4	$3\frac{1}{2}$,34
1=	0	1	5 ,78
	2	17	$3\frac{3}{4}$,17	for Alverdiscott.

49 **HUISH.**

Index, 4=	1	18	$8\frac{1}{4}$,04	Index, 3=	0	12	$10\frac{3}{4}$,01
1=	0	1	5 ,78	2=	0	4	$3\frac{1}{2}$,34
0=	0	0	$5\frac{1}{2}$,92		−0	17	$2\frac{1}{4}$,35
	2	0	7 ,74				
−0	17	$2\frac{1}{4}$,35					
	1	3	$4\frac{3}{4}$,39	for Huish.			

Reading the values from the chart on page xvi, one adds the values for indices 7 and 0 and then subtracts the values for indices 6, 5 and 2. The result is £28/16s/6d with a remainder of 13/100ths of a farthing. Note that no multiplication or division is needed, and that the calculations are all performed directly in £sd.

How much of a simplification is this new system? The advantages are greater as the number of parishes in the Poor Law Union increases. Fowler's system adds the step of calculating the table shown on page xvi, but once the table is made, the individual establishment charges can be calculated by simple addition and subtraction. Also, the results are already in £sd, so no additional conversion from farthings is needed.

For Fowler's example of a Poor Law Union with 23 parishes, his ternary system would require:
– 9 conversions to/from £sd to/from farthings
– 2 divisions of large values
– 6 divisions by 3
– 2 multiplications by 3
– 23 sums (adding and subtracting up to seven terms, once per parish)

Compared to the same example in decimal:
– 48 conversions to/from £sd to/from farthings
– 23 divisions of large values
– 23 multiplications of large values

Appendix 4

Fowler's Binary and Ternary Tables

David Hogan

The first positive step towards a reconstruction came with Hugh Fowler's reference to his father's binary and ternary tables published in 1838, designed to simplify the calculations he was required to make as treasurer to the Great Torrington Poor Law Union. The result, as Fowler himself says, of 'constantly searching after some other method more simple, and of easier application' was the discovery that 'any number might be produced by a combination of the powers of two or three.' At a time when decimal was firmly established as the number system of choice Fowler took a different view, proposing the use of a low order numbering system to simplify the process of calculation.

Fowler explored both the binary and ternary scales. The first nine terms of a binary, ternary and decimal series are:

100,000,000	10,000,000	1,000,000	100,000	10,000	1,000	100	10	1	Decimal
6,561	2,187	729	243	81	27	9	3	1	Ternary
256	128	64	32	16	8	4	2	1	Binary

For example, the decimal value 628 is expressed in binary as:

2^9	2^8	2^7	2^6	2^5	2^4	2^3	2^2	2^1	2^0
512	256	128	64	32	16	8	4	2	1
1	0	0	1	1	1	0	1	0	0

or

1x512	0x256	0x128	1x64	1x32	1x16	0x8	1x4	0x2	0x1

... and in ternary the decimal value 628 is expressed as:

3^5	3^4	3^3	3^2	3^1	3^0
243	81	27	9	3	1
2	1	2	0	2	1

or

2x243	1x81	2x27	0x9	2x3	1x1

The second stage of Fowler's revelation was that he could produce any number by the use of the indices of the ternary scale alone rather than the values that the ternary scale might represent. How Fowler achieved this can be seen in the later section entitled 'Fowler's Ternary Arithmetic'.

As far as is evident from Fowler's works and surviving papers, it seems that he believed his idea to be original. However, historians would point to the works of Vedic Mathematics, Fibonacci and Bachet as earlier examples of the application of signed number systems. Nevertheless, by choosing the ternary scale Fowler was able to exploit the most efficient of the scales available to him. Used in this way, the ternary scale enables representation of any value using the least number of elements. Whilst possibly a new notion to Fowler, mathematicians have been aware of this peculiar property of the ternary scale for centuries. Claude Bachet (1587-1638) challenged his peers to identify what minimum combination of weights for a two-arm weighing balance would be required to enable the measurement of any weight up to forty. The answer is of course, the ternary scale, or at least the first part of it: 1, 3, 9, and 27.

Using just the indices of the powers of 3, Fowler compiled tables that provided him with a significant short cut in each and every Poor Law calculation that he made. Or as he rather more poetically puts it, 'I was much delighted at observing the beautiful order in which the Indices introduced themselves into the Tables, and which made their construction a matter of great facility . . . '

Fowler's Ternary Notation

However, not content to rest on his laurels, Fowler concluded the introduction to his new tables with a hint that he was already working on another development:

> In the course of my observations on the Binary and Ternary Scales, I have fallen on a species of Binary and Ternary Arithmetic, which appears to possess some curious properties, but, as writing on this subject is incompatible with my present purpose, I have only given a short Example of Multiplication, in what may be termed Ternary Arithmetic, the process is extremely easy, and may be extended to very large numbers...

Fowler then goes on to demonstrate how two numbers, 628 and 564, are multiplied together using his ternary notation. The content of this table and the layout style is one Fowler uses as an illustration. The titles 'indices' and 'decimal' have been added for clarity.

indices	12	11	10	9	8	7	6	5	4	3	2	1	0		decimal
Multiply							+	0	-	-	+	-	+	=	628
by							+	-	+	0	0	-	0	=	564
						-	0	+	+	-	+	-	0	=	-1884
			+	0	-	-	+	-	+	0	0			=	50868
		-	0	+	+	-	+	-						=	-152604
	+	0	-	-	+	-	+							=	457812
	+	-	0	0	0	0	0	0	-	-	+	-	0	=	354192

Figure 5. Fowler's ternary notation showing 628 multiplied by 564. (Courtesy of David Hogan; data derived from an example in Fowler's papers.)

According to Fowler, this example 'scarcely requires any mental exertion whatever; no multiplication, nor even addition, is required, as ordinarily practised.' At this stage the workings are probably incomprehensible. But not when the logical development of Fowler's work on ternary is more fully explained.

The notation used in the example uses a convention Fowler proposed and that has come to be known amongst the project team as "Fowler's Ternary Logic." The underlying number system is more usually referred to as "signed ternary" or "balanced ternary". Fowler's notation is fundamentally simple. He placed a plus sign (+) in the ternary scale column when the particular index value was to be added, a minus sign (−) when it was to be subtracted, and zero (0) when neither was required. So using Fowler's ternary notation we now have another way of representing the decimal number 628.

3^5	3^4	3^3	3^2	3^1	3^0
243	81	27	9	3	1
2	1	2	0	2	1

or

2x243	1x81	2x27	0x9	2x3	1x1

Knowing that this is a ternary table, the column headings now become superfluous. Plus and minus acquire a value of their own because of the column position they occupy. I.e. where a plus occurs the value of the indices is added, where a minus occurs the value of the indices is subtracted — unlike binary where 1 indicates the value of the indices is present and 0 indicates that it is not present.

| + | 0 | - | - | + | - | + | = 628 |

| + | - | + | 0 | 0 | - | 0 | = 564 |

Figure 6. Plus and minus acquire a value of their own simply because of the column position they occupy. (Courtesy of David Hogan.)

It is then just a matter of applying a few simple rules to complete any arithmetical operation using these symbols. Once these are mastered, Fowler's short example of multiplication does indeed become as simple as he claims.

Thomas Fowler's Ternary arithmetic

Using Fowler's ternary notation, his rules for addition, subtraction, multiplication and division are as follows.

Addition

Example: To add 36, 14 and 30

Plus added to minus yields zero — that is, cancelled

Plus added to plus yields minus, plus being carried to the next column on the left.

Plus added to plus added to plus yields zero, plus being carried to the next column on the left.

Minus added to minus yields plus, minus being carried to the next column on the left.

+	+	0	0		=	36
	+	-	-	-	=	14
	+	0	+	0	=	30
+	0	0	0	-	=	80

Subtraction

Example: To subtract 35 from 54

Invert the signs of the subtrahend and add, using the rules for addition previously given, to the line to be subtracted from.

+	-	0	0	0	=	54
		-	-	0	+	=	-35
0	+	-	0	+	=	19

Multiplication

As with familiar decimal multiplication, each digit of the multiplier is separately multiplied by each digit of the multiplicand, the final result being the summation of the partial products. In Fowler's ternary system, in the formation of each separate product, for every plus in the multiplier bring down the corresponding sign in the multiplicand. In the example 23 x 19, the sum of the three partial products is found using the rules of addition as previously given.

			+	0	-	-	=	**23 multiplicand**
			+	-	0	+	=	**19 multiplier**
			+	0	-	-		**23 partial product**
		0	0	0	0			**0 partial product**
	-	0	+	+				**-207 partial product**
+	0	-	-					**621 partial product**
+	-	-	+	+	-	-	=	**437**
729	**243**	**81**	**27**	**9**	**3**	**1**		**decimal equivalents**

Division

Arrange as for common long division. However, the signs of the terms of the divisor must be inverted — that is, plus becomes minus, minus becomes plus with zero maintained as zero.

8	280	35
divisor	dividend	quotient
+ 0 -		
inverted		

-	0	+)	+	0	+	+	0	+	(+ + 0 -
			-	0	+				
				+	-	+			
				-	0	+			
						-	0	+	
						+	0	-	
						0	0	0	

    Select an appropriate term (sign) for the quotient so that when multiplied, with the inverted divisor (using the rules for multiplication above) the resulting product when added to the first corresponding terms of the divide, reduces the first term or terms of the divide to zero and so on. See http://www.mortati.com/glusker for a detailed explanation.

Once these rules are mastered, Fowler's short example of multiplication does indeed become as simple as he claims.

indices	12	11	10	9	8	7	6	5	4	3	2	1	0	decimal
Multiply							+	0	-	-	+	-	+	= 628
by							+	-	+	0	0	-	0	= 564
						-	0	+	+	-	+	-	0	= -1884
			+	0	-	-	+	-	+	0	0			= 50868
		-	0	+	+	-	+	-						= -152604
	+	0	-	-	+	-	+							= 457812
	+	-	0	0	0	0	0	0	-	-	+	-	0	= 354192

Figure 5. Fowler's ternary notation showing 628 multiplied by 564. (Courtesy of David Hogan; data derived from an example in Fowler's papers.)

There is one particular feature of this form of ternary that eases the handling of negative numbers. With his use of minus, zero, and plus in any one column position that is, using the indices of the ternary scale, instead of the more usual ternary digits of 0, 1, 2, Fowler found that handling negative numbers became as easy as handling positive ones. The most significant digit, the far-left-hand one, of any number represented in this form of ternary gives the clue. If this is plus (+), as it is with both our examples of 628 and 564, then the whole value is positive. If the most significant digit is minus (−), the whole value will be negative since the sum of the less significant digits, even if all these are positive, will always be less than the most significant digit. For example the ternary value expressed, in most significant to least significant digit order, as (−, 0, +) tells us to subtract the 9, ignore 3 and add 1 giving the answer −8, a negative number. [(−1 x 9) + (0 x 3) + (1 x 1) = −8.]

This property of signed or balanced ternary is of particular interest to computer science. Conventionally, digital computers using the binary system struggle when required to represent negative values. The standard solution involved the method of two's complements. To represent a negative number using two's complements, the positive binary number is inverted digit for digit, and a 1 is added to the result. The practical difficulty of the technique is that it requires a number of extra steps in every calculation involving negative values. Effective but cumbersome.

The process required to represent −28 in two's complement form:

	0	0	0	1	1	1	0	0	= 28
invert	1	1	1	0	0	0	1	1	
add 1	1	1	1	0	0	1	0	0	= -28

Figure 7. Representing –28 in two's complement notation. (Courtesy of David Hogan.)

How much simpler it is to represent a minus value in signed, or balanced, ternary.

So far our investigation into Fowler's ternary logic has used positive or negative integers as examples. The system is easily extended from integers to real numbers. As in the decimal system, where a decimal point discriminates between the whole and the fractional components of the number, similarly in Fowler's signed ternary notation a "tertial" discriminator is employed for the same purpose. In Fowler's notation this takes the form of a vertical line.

3^3	3^2	3^1	3^0	3^{-1}	3^{-2}	3^{-3}		
+	-	+	+				=	22.00
	+	-	+	+			=	7.33
		+	-	+	+		=	2.44
			+	-	+	+	=	0.81

For a more complete analysis of the method see IEEE Annals of the History of Computing 27:3 (July/September 2005): 4–22.

Appendix 5

Description of a calculating machine, invented by Mr. Thomas Fowler, of Torrington in Devonshire. By Augustus De Morgan, Esq. Communicated by Francis Baily, Esq. V.P.R.S.

By permission of the Royal Society. Ref: AP 23.24

Mr. Fowler had been employed by the Guardians of the poor of the district in which he lives to calculate the manner in which a given total assessment should be levied upon the different parishes of the union, each parish being rated in proportion to a given sum. When the sums which express the proportions of the ratings are fractional, such a process, though simple enough in principle, is complicated in its operation, if it be desired to divide the assessment with legal precision.

It suggested itself to Mr. Fowler that the process of multiplication assumes a simplicity in the binary and ternary systems of arithmetic, which might make it desirable to try whether it might not be worth while to convert the factors into one of these systems by tables, to perform the operation upon the converted results, and to reconvert the product into the common system by an inverse use of the table. The binary system appearing too tedious, the ternary was preferred. In this system every digit must be either 0, 1, or 2, but if a positive and negative unit be employed, every digit is either -1, 0, or $+1$, say $\bar{1}$, 0, or 1. Thus the number which decimally expressed is 70, becomes in the ternary system 2 1 2 1, and 2 being 1 $\bar{1}$, this may be made 1 0 $\bar{1}$ $\bar{1}$ 1.

Since unity is in multiplication only an index, it is obvious that the rules of multiplication and division, when the system is ternary, and the positive and negative units are employed, must consist entirely in directions for the management of the signs + and −. The accompanying paper, by Mr. Fowler, contains instances.

Mr. Fowler, having drawn up and published some of the necessary tables, and having found this method convenient in practice, began to consider whether the mechanical part of the calculation might not performed by machinery. The instrument which he caused to be constructed by the workmanship of a country carpenter, though large and difficult to move, is easily used. The following is such an account of it as can be given without drawings, and Mr. Fowler is occupied in preparing one of a more detailed and complete kind.

The machine consists of four essentially distinct parts. The first, second and third exhibit the multiplicand, multiplier and product, or quotient, divisor and dividend, according as the question to be worked is one of multiplication or division. The fourth is a carrying apparatus, which though at present detached, and employed to reduce the result to its simplest form after the main operation has been performed, might without much difficulty be attached to the multiplier or divisor, and work with it.

Let us now suppose a question of multiplication, both multiplier and multiplicand being exhibited in the ternary system. The multiplicand consists of nothing but a number of rods, each bearing an index, and each movable backwards and forwards. When the indices are all arranged in line, and in one particular line, the multiplicand is 000.... But if any one of the rods be advanced by a certain space forwards, the digit +1 is indicated as occupying the numeral column which that rod represents; and if it be moved the same space backwards, −1 is the digit indicated. This, which we may call the frame of the multiplicand, is thus a collection of rods, not itself connected with any machinery, but only useful as indicating the manner in which the frame of the multiplier is to act.

The multiplier is a frame movable in the direction perpendicular to the rods of the multiplicand and product, and situated between the planes of the two, in such manner that its extremity can be brought by a sliding motion over each rod of the multiplicand in succession. This multiplier consists of a number of rods in a common system, each furnished with two teeth, one at each extremity, the tooth by which it is acted on being a continuation of the rod, that by which it acts being perpendicular to the axis of the rod. The first set of teeth are disposed so as to rest in a frame which has a slight motion round an axis; and each rod can be moved so that its teeth shall touch the frame, above, on or below, the axis. Those rods, then, which have their teeth on the axis do not receive motion from the frame, while the others receive motion in one direction or another, according as the teeth touch the frame above or below the axis. The perpendicular teeth at the other extremities may thus be made to move in either direction, or to remain

174

stationary: and these last mentioned teeth act upon the rods which make up the frame of the product. This last frame precisely resembles the frame of the multiplicand, with the addition of the connecting part by which the multiplier acts upon it.

The process of multiplication is then as follows; the frame of the multiplier having been set, and also that of the multiplicand, the extremity of the multiplier frame is brought over the first rod of the multiplicand. To this extremity is attached a tooth which acts upon the rod of the multiplicand over which it comes, giving it a motion in one or the other direction, according as the slightly revolving frame of the multiplier is made to move in one or the other direction. The rule is, to move the revolving frame in such a way as to bring the rod of the multiplicand to its zero position: and this one motion multiplies the figure of the multiplicand by the whole of the multiplier, and by the action of the perpendicular teeth, exhibits the result upon the product-frame. The lateral motion is then given to the whole of the multiplier apparatus, until the tooth comes upon the next figure rod of the multiplicand, and the revolving frame being then made to bring the new multiplicand rod to zero, the effect upon the product frame is that the new figure of the multiplicand is multiplied by the whole of the multiplier, and the result added to that of the preceding figure. This process is continued until the whole of the figures of the multiplicand are exhausted.

The result is then completely exhibited on the frame of the product, but not in its simplest form. For, whereas $+1$ or -1 should be the only digits in the final result, this intermediate result may exhibit $+2$ or -2, $+3$ or -3 etc on any rod. The carrying frame is a simple apparatus which, like the multiplier, has a lateral motion, and can be brought on any pair of consecutive rods. By one motion of the hand, it advances the left of two rods by a unit, and throws back the right hand rod by 3 units, or vice versa. Some little expertness is here necessary in making the carriages properly, with reference to the simplicity of the result: but there is no possibility of absolute error being introduced, since each process can only consist in altering a lower column by 3 units while the next column is altered in the contrary way by one unit.

The method of performing division is precisely the reverse of the preceding, and will hardly need description.

This machine, which was lately in London, was exhibited to many gentlemen, Fellows of the Society and others, whose united opinion on the ingenuity of the contrivance suggested to the friends of Mr. Fowler that a slight description might not be unacceptable to the Royal Society.

The instrument confessedly does not realise the notion of a calculating machine properly so called, since the necessity for using tables both for converting the factors, and reconverting the result, introduces both labor and risk of error.

It is also felt that the preceding description is insufficient to give more than a mere glimpse of the principle and detail. Such as it is, however, the Society will perhaps not regret the bestowal of one moment of its attention, when it considers that the inventor will thereby not only feel honored and gratified but secure in the possession of the credit to which his ingenuity is entitled, and for which alone he had labored.

Appendix 6

Thomas Fowler on his calculating machine. General Notation, Notation for the Calculating Machine, Tables

Reproduced by permission of the Royal Society Ref: AP 23.26

and any of the other Coefficients either 2.1. or 0, and as the Principle of this Calculating Machine admits of only two Coefficients namely 1, & 0, a substitution for the Coefficient, 2, must be made whenever it occurs in any number in the common Ternary Notation.— This may be done in the following manner, which also gives the Law of Continuation for a Table for consecutive numbers carried onwards to any extent whatever.

Taking the Ternary Series $N = a_z 5^n + a_z 5^{n-1} + \&c. + q.3^3 + a_z 5^3 + q.3^3 + a_z 5^4 + a_z 5^1 + a_z 5^2 + a_z 5^4$ &c. and calling the Terms, commencing on each side of the Initial Line, the $1^{st} 2^{nd} 3^{rd} 4$ &c. Terms to the left or right of the Initial Line, and making the consecutive substitutions of 0, 1 & 2 for each of the Coefficients a, a, q &c. the various combinations arising, will give the following Tables— the figures substituted for the Coefficients on the left side make the common Ternary Notation and the Signs on the right side make the Notation used for the Machine & their value in the Ternary Notation is placed in the Column adjoining.——

a_z	a_z	a_z	a_z			

(A large handwritten table of ternary-notation substitutions and sign combinations, numbered 1 through 48 in the right-hand column.)

In this small Table which is continued through all the combinations of the Coefficients $q, q, q, 4q$, the law of continuation is manifest, the lines of signs are called Ternary Numbers, and the signs $+, -, 0,$ proceed downwards in Groups of the Powers of 3 according to their Order of place in the Ternary Notation, and when $-$, (the sign ellipsis) commences in the first Group of $-$, in any Order the first $+$ commences in the same line in the next higher or superior Order. —— Any Positive Ternary Number is made Negative by changing all the signs composing it. viz. by changing $+$ into $-$, and $-$ into $+$, and conversely any Negative Number is made Positive by changing all the signs in the same manner, in such cases these Numbers whether Positive or Negative are said to be inverted. It is also evident that $+$ and $-$, in any the same Order cancel each other or are $= 0$. that is $\frac{+1}{-1} = 0$. which is the same as saying $+3^- 3^- = 0$. —— Any Number shifted one Term or degree to the right over a Tertial Line is reduced to One Third of its original value, if it be shifted One Term to the left and 0 added in the Unit Term's place, if it had no Tertial, its value will be increased threefold, and so on in proportion as in common Arithmetic.

A Tertial Line is always supposed to stand to the right of the Unit Term in any whole Number, and to the left of the first Term if a Tertial Fraction. —

If the Ternary Number $+-++ = 22$ have its Units Term cut off by a Tertial line it will take the form $+-+|+ = 3^2-3+5^1+5^{-1}$, it is thus reduced to One Third of its first value being now only $7\frac{1}{3}$. if the Tertial line cut off two Terms it will be reduced to One ninth of its first value namely $+-|++ = 2\frac{4}{9}$ $= 3^1-3^0|+5^{-1}+5^{-2}$, the same may be seen by looking for these expressions in the above Table, the Tertials being the Numerators, to corresponding Powers of 3 for their Denominators according to the number of Terms in the Tertial; If the Tertial line cut off three Terms, thus $+|-++ = 3^0|-5^{-1}+5^{-2}+5^{-3}$, the number $+-++ = 22$ is reduced to one twenty seventh of its original value; in this case the Tertial $|-++$ is a negative Fraction and is $= -\frac{8}{27}$, the numerator of which may be found in the above Table by looking for the Signs of the Tertial when inverted. the value of the whole expression $+|-++$ is therefore $1-\frac{8}{27} = \frac{19}{27}$. —— In cases, when the value of a number having a Negative Tertial, is required in the common notation, it is very convenient to make the Tertial Positive, otherwise a troublesome subtraction is sometimes necessary. —— In the Ternary Notation, the local value of every Unit being increased threefold, in the next Order above it to the left, therefore $+$ or $-$ in any Order are equal to $3+$ or $3-$ in the next Order below them to the right respectively, and as $-$ cancels $+$ in the same order so it would cancel Three or a triple plus, $\mp$, in the next Order below it, that is $-\mp = 0$. consequently $-|\mp = 0$, which is the same as saying $-3^1+3\times5^0 = 0$, or $-3^0+3\times5^{-1} = 0$, this expression $-\mp$, may therefore be added to any two consecutive Terms in a Ternary number without increasing or diminishing its values. — In the above expression $+|-++$ if $-|\mp$ be placed under the corresponding Terms thus $\frac{+|-++}{-|\mp}$, and be added to them the $+$ and $+$ cancel each other and the $-$ in the Tertial cancels one of the $+$ in the $\mp$ and the whole expression becomes $|\mp++$ a Tertial Fraction $= \frac{2\times5^{-1}+5^{-2}+5^{-3}}{}$ $\frac{2\times3^1+3^0}{3^3} = \frac{19}{27}$ as before. —— Again in the mixed number $+0-0|00-+ = 24\frac{10}{27}$ by adding $-|\mp$ in the Tertial place it becomes $+0--|\mp0-+ = 23\frac{37}{27} = 3^2-3^1-3^0|+3\times5^{-1}-5^{-2}+5^{-3}\frac{23\times27-23}{27}=\frac{621}{27}$ &c &c &c this mode of using the expression $-|\mp = 0$, will be found very useful with the calculating Machines, it is only, the Borrowing One Unit from the whole number, & incorporating it with the Minus Tertial.

In this way any whole or mixed number of any magnitude whatever may be accurately expressed by $+1, -1$ and 0, and if $+$ and $-$ be understood to represent positive & negative Units, and their places in any representation of number, the places of the Orders of Powers of 3 as they stand in this representation of Ternary Numbers, the work of Addition Subtraction Multiplication & Division becomes merely an Algebraical Operation, with the Signs $+, -$, and 0, only, and it is also exceedingly well adapted for Mechanical Operation.

Examples

Addition

Add $+-0+$	$= 19 = 3^3-3^2+5^0$	
To $++0-$	$= 35 = 3^3+3^2+3^0$	
Sum $+-000$	$= 54 = 3^3\cdot3^0 = 2\times3^3 = (3^1-1)\times3^3$ &c &c &c	

The Rules for Addition can easily be inferred from the foregoing.—namely $+$ & $-$ in the same Order $=0$ *or cancel each other*
Three $+$ in the same column make 0, in that column, and carry One $+$ to the next column or Order to the left, Three $-$ in one column make 0, in that column and carry One $-$ to the next column to the left Two $+$ in one column, make $-$, in that column and carry One $+$ to the next column to the left Two $-$ in one column make One $+$ in that column and carry One $-$ to the next column to the left.

In the first column in the above example + & − make 0, in the second column 0, make 0, in the third Column + & − cancel each other, and in the fourth Column Two + make − in that Column and carry + to the place of the 5th Order.

$$\begin{array}{r} Add\ ++\ 00 = 36 \\ +---\ = 14 \\ And\ +\ 0+0\ = 30 \\ \hline +000-\ = 80 = 3^2-3^0 = 81-1 \end{array}$$

The same Rules are observed in this Example the 3+ in the 4th Column make 0, in that Col. & + for the 5 place

$$\begin{array}{r} Add\ ++0+ = 37 \\ +++0 = 39 \\ And\ +-++ = 22 \\ \hline ++-0- = 98 = 3^4+3^3-3^2-3^0. \end{array}$$

Here 2+ are − and + for the next Term, where 3+ are 0 and + for the 3 term, where + & − cancel each other and 2+ are − & + for the 4th Term, in this term + & 3+ make one + and one + for the 5th place

$$\begin{array}{r} Add\ +---\ = 14 \\ +0-- \ = 23 \\ And\ ++-0\ = 33 \\ \hline +0--+ = 70 = 3^4-3^3-3^1+3^0 \end{array}$$

Here 2− are = + and − for the second Term, this with 3− make One − and One − for the third Term in the third Term, the − brought to this Term cancels the +, and the − is brought down &c

$$\begin{array}{r} Add\ +0\ \ -+ \ = 3-\tfrac{8}{9} \\ +\ -- = 1-\tfrac{4}{9} \\ +\ +0 = 4+\tfrac{3}{9}\ or\ \tfrac{1}{3} \\ \hline +0\ -0\ = 8-\tfrac{1}{9} = 3^2-3^0-3^{-1}\ a\ Negative\ Textial \\ Add\ -\ \mp 0\ = 0 \\ \hline +-+\ \mp 0\ = 7\tfrac{8}{9} = 7\tfrac{8}{9} = 3^2-3^1+3^0+(2\times3^{-1})\ a\ Positive\ Textial \end{array}$$

Here − & + in the lowest Term of the Textials to the right cancel each other, + & − in the next terms cancel each, and the remaining − is brought down, the Columns in the whole numbers are added as before, and − 1⊧ is added, to make the negative Textial positive.

$$\begin{array}{r} Lastly\ Add\ ++0+\ +-+0 = 37\tfrac{8}{9} \\ +--\ \ +++ = 14\tfrac{8}{9} \\ ++0-\ 000+ = 35\tfrac{1}{9} \\ \hline +0+-0\ +0- = 87-\tfrac{1}{9} = 3^4+3^1-3^0-3^{-1}+3^{-2}-3^{-4} \\ Add\ \ -\ \mp000 = 0 \\ \hline +0+--\ +0- = 86\tfrac{2}{9} = 3^4+3^1-3^0-3^1+2\times3^{-1}+3^{-2}-3^{-4} \end{array}$$

In this Example, One + is carried from the first Column of Textials to the whole Number, the rest as before, & the whole is proved by the binary Notation.

Subtraction

Rule. Invert or change all the Signs in the Line to be subtracted, and add the Two Lines together

$$\begin{array}{r} From\ +-000\ = 54\ take\ ++0-\ = 35 \\ ++0-\ Inverted\ +\ \ ++0+\ =-35 \\ \hline Difference\ +-0+\ = 19 \quad see\ 1^{st}\ Example\ above\ in\ addition \end{array}$$

Multiplication

Rule. For every + in the Multiplier bring down the Multiplicand to its proper place and for every − in the Multiplier bring down the Multiplicand inverted to its proper place, according to the orders of the Terms in the Multiplier the sum of the whole is the Product

$$\begin{array}{r} Multiply\ ++0-\ =35 \\ By\ \quad +0-\ = 8 \\ --0+ \\ ++0-0 \\ \hline +0++0+ = 280 = 3^5+3^3+3^2+3^0 \end{array}$$

Here − in the first Order of Terms in the Multiplier brings down the Multiplicand inverted to its proper place in the operation and +, in the third Order of Terms in the Multiplier merely brings down the Multiplicand to its proper place in the third Order,

$$\text{Multiply} +0-- = 23$$
$$\text{by} \quad +-0+ = 19$$
$$+0--\quad 207$$
$$-0++0\quad 23$$
$$+0--$$
$$\overline{+--++--} = 437 = 3^6 - 3^5 - 3^4 - 3^3 + 3^2 + 3^2 - 3^1 - 3^0}$$

This operation is similar to the foregoing and does not require any explanation whatever

$$\text{Multiply} +0-- = 3-\tfrac{2}{3} \quad \text{a mixed number}$$
$$\text{by} \quad +-|0+ = 2+\tfrac{1}{3} \quad \text{a mixed number}$$
$$+0--$$
$$-0++0$$
$$+0--$$
$$\text{Product} = \overline{+--++--} = 5\tfrac{11}{27}$$

In this Example the Tertial in the Product contains four terms this being the sum of the Terms of the Tertials in the Multiplicand & Multiplier as common Multiplication of Decimals

Division

Rule, Arrange the Dividend, and the Divisor inverted as in common Division, find a sign or Term for the Quotient, which multiplied with the inverted Divisor and the product added to the first corresponding Terms of the Dividend reduces the first term or terms of the Dividend to Zero or 0, take down the next sign and proceed in the same manner to the end, or as far as needful by adding 0s if Tertials are required.

$$\text{Divide} +0++0+ = 210 \text{ by } +0-- = 8$$
$$\text{The Inverted Divisor} -0+) +0++0+ (++0- = 35 \text{ the Quotient}$$
$$-0+$$
$$+-+$$
$$-0+$$
$$-0+$$
$$+0-$$

By this Example the inverted Divisor is -0+ it being the given Divisor +0- with the signs changed the first terms as they arise in the Quotient are indicated by the first terms of the remainders or rather sums, as the Divisor is inverted and added, which is the same as subtracting the proper Divisor, —

$$\text{Divide} +--++-- = 437 \text{ by } +-0+ = 19$$
$$\text{The inverted Divisor} -+0-) +--++-- (+0-- = 23 \text{ the Quotient}$$
$$-+0-$$
$$-0+-$$
$$+-0+$$
$$-+0-$$
$$(0\ 000)$$

If the Divisor in this Example be made a mixed number having two Terms in the Tertial thus +-|0+ = 2⅓ and the Dividend remain the same Two 0s must be added to the Quotient making +0--00 = 207 = 3⁵-3⁴-3³ &c &c &c. —

In this way any Arithmetical Operations, whether simple or compound, may be accurately performed and the whole is applicable to Mechanical Operation with little expense or trouble —

Thos Fowler June 3 1840

Appendix 7

Notice of Mr. Fowler's new Calculating Machine. Communicated by Professor Airy

©The British Library Ref: AC 1181 Transactions of the British Association for the Advancement of Science. Vol 10. September 23, 1840

The origin of this machine was to facilitate the calculation of the proportions in which the several divisions of a poor-law district in Devonshire were to be assessed. The chief peculiarity of the machine is, that instead of our common decimal notation of numbers, a ternary notation is used; the digits becoming not tenfold but threefold more valuable as they were placed to the left; thus, 1 and 2 expressed one and two as in common, but 1 0 expressed (not ten, but) three, 1 1, four, 1 2, five; but again 2 can be expressed by 3, with 1 taken from it. Now, let $\bar{1}$, written thus with a small bar above it, mean that it is subtracting; then 1 2 and 2 $\bar{1}$ are the same in effect, both meaning five; and, for a similar reason, replacing 2 by its equivalent 1 $\bar{1}$, we have five written in three several ways: 1 2, or 2 $\bar{1}$, or 1 $\bar{1}$ $\bar{1}$; the last is the form used. It is obvious, that by an assemblage of unit digits thus positively or negatively written, any number may be expressed. Thus the number which decimally expressed is 70, becomes in the ternary system 2121: and 2 being equal to 1 $\bar{1}$, this may be made 1 0 $\bar{1}$ $\bar{1}$ 1. In the machine, levers were contrived to bring forward the digits 1 or $\bar{1}$, as they were required in the process of calculation. A full description of the machine, drawn up by Professor De Morgan, was presented by the author.

Appendix 8

Sir Trevor Wheler/G.B.Airy correspondence

By permission of the Science and Technology Facilities Council and the Syndics of Cambridge University Library.
Ref: Cambridge University library, Royal Greenwich Observatory Archive. RGO 6/427 52

Ilfracombe May 13, 1841
Sir, I received a letter some time ago from Sir John Forbes enclosing the copy of a note from yourself to Professor Forbes on the subject of Mr. Fowler's Calculating Machine, a notice of which I believe you were kind enough to insert in the Athenaeum of October last. I was in hopes Mr. Fowler might have been able to furnish some Plans & Sections of his Machine, but he tells me he is not prepared to offer any further details of its construction than those which have already appeared in a Paper drawn up by Mr. De Morgan & read, I believe, at the meeting of the British Association last year. He has however requested me to forward the enclosed letter, & I venture to accompany it with an attempt to explain in some degrees the operation of the instrument by shewing the manner in which the Addition, Subtraction and Multiplication of the Signs used by Mr. Fowler to express his numbers is performed on paper, since it represents exactly the process that takes place in the Machine itself: for nothing can be more simple than its construction. Division is performed with equal facility by the Machine, but cannot be so readily explained on paper.

Mr. Fowler is in great hopes of being able to exhibit his invention at Devonport in the autumn, but should he be prevented & you will do me the favour of taking my residence "Cross House, Nr Torrington" in your way either to or from the Meeting of the British Association, it will afford me very great pleasure to be honored by your Company
I remain, Yr most obedt. sevt. Trevor Wheler

Addition:

When there is an equal number of Plus and Minus signs in a term they cancel one another & 0 is put down below. But when they are unequal and one Plus or one Minus is over, it is put down underneath. If two Plus Signs are over, put down a Minus & carry a Plus to the next superior term: for 2=3−1 or +−. On the same principle if two Minus Signs are over, put down a Plus & carry a Minus to the next term. When three Plus or three Minus Signs are over carry a Plus or a Minus to the next term & put down 0. And so on, carrying a Plus or Minus to the next superior term for every three in the same term.

Example - add together 445,148 & 126

Add together 445, 148, & 126

Power	6th	5th	4th	3rd	2nd	1st	0 or unit				
	+	-	-	+	+	+	+	represents	445	or	$(729+27+9+3+1) - (243+81)$
		+	-	-	+	+	+	"	148	or	$(243+9+3+1) - (81+27)$
		+	-	-	-	0	0	"	126	or	$(243) - (81+27+9)$
	+	0	0	0	-	0	-	"	719	or	$729 - (9+1)$

N.B. The first operation of the Machine exhibits the excess of Plus or Minus signs in each column thus:

$$+ +$$
$$+ + \; - \equiv \; - + + +$$

and the "Carrying Frame" reduces them, according to the Rule laid down, to the formulae in the answer.

Subtraction: Invert the signs corresponding to the lesser number & add them to those of the greater.

Example: required the difference of 2770 & 928.

Required the difference of 2770 & 928

+	+	-	+	+	-	-	+	represents	2770
	+	+	-	+	+	0	+	"	928

then

+	+	-	+	+	-	-	+		
	-	-	+	-	-	0	-	Inverted signs of 928	

Answer | + | 0 | - | - | - | + | - | 0 | | 1842

1st operation of machine exhibits

Multiplication
        Place the signs of the Multiplier beneath those of Multiplicand & begin with the Unit sign of the Multiplier as in common Arithmetic. If it is a Plus sign put down the Signs of the Multiplicand as they stand, but if it is a Minus Sign invert them. And so on with the next.

Example Required the product of 866 & 16.

Required the product of 866 & 16

		+	+	-	-	0	+	-	represents	866
			+	-	-	+			"	16
		+	+	-	-	0	+	-		
	-	-	+	+	0	-	+			
	-	-	+	+	0	-	+			
+	+	-	-	0	+	-				

Answer | + | - | 0 | + | 0 | 0 | 0 | + | - | - | | 13856

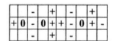

1st operation of Machine exhibits

185

Appendix 9

Thomas Fowler\G. B. Airy correspondence

By permission of the Science and Technology Facilities Council and the Syndics of Cambridge University Library.
Ref: University of Cambridge Library, Royal Greenwich Observatory Archive RGO 6/427 f54r

Dear Sir

My friends and patrons here, among whom I would particularly name Sir Trevor Wheler, are anxious that I should thank you for your notice of my new calculating machine at the last meeting of the British Association, as reported in the Athenaeum of 31 October last. Please to accept my sincere and hearty thanks for this honour and if I may be permitted to intrude on your valuable time, I would give some further account of this machine, which perhaps may hereafter be found valuable as well in small as in the most extensive Arithmetical Calculations; The extent of the principle is only limited by the means to which human power can be applied mechanically, and the Machine I now have, which contains 55 Terms or places in the Ternary Scale enables me to perform Arithmetical Operations with factors or divisors not exceeding 28 of those Terms or above Ten Million Millions and downwards to the one ten million millionths part of a Unit, which may be almost instantly be made to operate on any Number (whole or mixed) from 87 Quadrillions down to the One Thousand Quadrillionth part of a unit; and in regard to a Quotient or Reciprocal, the divisions can be carried on indefinitely. This machine was constructed entirely with my own hands (principally in Wood) with the utmost regard to Economy, and merely to put my ideas of this mode of calculation into some form of action; It is almost six feet long, one foot deep and three feet wide, – In brass & Iron it might be constructed so as not to occupy a space much larger than a good portable writing desk and with powers such as I have described – – – –

I had the honour in May 1840 to submit the Machine to the inspection of many Learned Men in London among whom were the Marquess of Northampton, Mr. Babbage, Mr. F. Baily and A. de

Morgan, Esq. with many other Noblemen and Gentlemen Fellows of the Royal Society etc and it would have been a great satisfaction to me if I could have had the advantage of your opinion also. They all spoke favourably of my Invention but my greatest wish was to have had a thorough investigation of the whole principle of the Machine and its details as far as I could then explain them, in a way very different from a popular exhibition, – this investigation I hope it will still have by some first rate Man of Science before it be laid aside or adopted. – I am fully aware of the tendency to overate one's own inventions and to attach undue importance to subjects that preoccupy the mind, but I venture to say and hope to be fully appreciated by a Gentleman of your scientific attainments, that I am often astonished at the beautiful result of a calculation entirely mechanical; I can easily understand your feelings at the (perhaps) unexpected result of a differential equation unravelled entirely by your own patience and ingenuity; such possibly, though of a lower grade, are my feelings at the results of my mechanical operations; conscious that the whole is as yet untrodden Ground, I would in illustration give one instance. – Since I left London I have applied the Machine occasionally to the calculation of Logarithms, and, from the comparative ease with which this may be done even to 70 or 80 or indeed to any number of places of figures, and, with the same facility either for the Logarithm or Natural Number I am almost inclined to think that this may properly be called a Logarithmic Machine. My present clumsy performance will give a Logarithm or Natural Number true to 26 or 27 places and the limit is only the extent of human mechanical agency.– – –

Wishing to calculate the Natural Number to Hyperbolic log.1, or the Base e of the Napierian System to 12 or 13 places, the Machine produced the Ternary expression, +0/–0+++. – –.0–0. –+++–. 0+++–. 000–+, 0– – – –, which to the right of the line a, at the beginning (called the Tertial line) is composed of Negative Powers of 3, according to their places from the Tertial Line and taken as positive & negative numbers as the signs represent and with the + 0, to the Left is .e accurately and may be used for all the purposes of mathematical calculation if the ternary notation be used; – The value of this or any other expression or … (ed. — unreadable) found or turned into the decimal Notation by Tables which I exhibited to Mr. Babbage and other Gentlemen during my late visit to London, which Tables in a

small compass will give any decimal accurately to 27 places of figures,
– The above is turned into the decimal Notation by those Tables in
the following manner premising that it is divided into Classes of 5
Terms each; and that the first term of any Class, if negative, will cause
the arithmetical Compliment of the decimal to be taken to make the
whole additive, and also that ‡ signifies "double Plus" or 3^{-1} x 2, in its
place to make the first Class of the Tertial for this example, additive
which consequently takes Unity from the Whole Number + 0 or 3, and
makes it +−, or 2 according to any adaption of the Ternary Scale.

I am strongly impressed with the conviction that this method of
calculating Logarithms may even now be found highly useful in
questions of a philosophical nature where a Logarithm to a
considerable extent is required. It is entirely independent of the
method of differences and proceeds to the Logarithm or Natural
Number at once, so that had I to extract a very high root of any
number such as the 365¼ths, or 1001⅞ths etc etc It would be better to
calculate the Hyperbolic Logarithm of the given Number for this
specific purpose and divide it by the Index of the Root etc etc than to
use any general Tables that can possibly exist of human construction,
and I am certain that the whole ternary expression for the required
result may be obtained mechanically by properly carrying out the
principle I have proposed, and I have no doubt whatever that men of
abilities far superior to my own will hereafter be found to improve this
principle so as to make it generally useful −−−−−
    I often reflect that had the Ternary instead of the denary Notation
been adopted in the Infancy of Society, Machines something like the
present would long ere this have been common, as the transition from

mental to mechanical calculation would have been so very obvious and simple. — — — —

I am very sorry that I cannot furnish any drawings of the Machine, but I hope I shall be able to exhibit it before the British Association at Devonport in August next where I venture to hope and believe I may again be favoured with your valuable assistance to bring it to notice, − I have led a very retired life in this Town without the advantage of any hints or assistance from any one, and I should be lost amidst the Crowd of Learned and distinguished Persons assembled at the Meeting, without some kind Friend to take me by the hand and protect me,− −

I remain Sir, with great respect

Your Obedient Servant,

Thos.Fowler

Great Torrington Devon May 8, 1841

To
   Professor Airy
      Royal Observatory
         Greenwich

Appendix 10

On a New Calculating Machine. By Mr. Fowler.

© The British Library Ref AC 1181 Ref: Transactions of the British Association for the Advancement of Science.
Vol. 11 1841

The machine itself is on the principle of the old abacus, or calculating rods. At the first glance it has somewhat the appearance of a pianoforte, or organ, with all its keys laid bare. It would be impossible by mere description to make clear the details of the instrument or the method of working it. The property of numbers on which Mr. Fowler bases the notation of his machine is, that any number whatever may be expressed by a proper combination of the powers of the number 3. The powers of 3 are in succession thus: the 0 power is 1; 1st, 3; 2nd, 9; 3rd, 27; 4th, 81; 5th, 243; &c. &c. Thus the number 14 may be expressed by subtracting from 27, or the 3rd power, the sum of 9, 3, and 1, the 2nd, 1st, and 0 powers, and similarly for all other numbers; the combination for some being more simple, for others more complicated. Instead of using the nine common characters with 0, nought, or zero, as in our common mode of numbering, Mr. Fowler only uses three marks, + when the power of 3 is to be added, − when it is to be subtracted; and the power itself is expressed by the place in which the mark stands: thus the number 14 would be + − − −, where the − most to the right means that the 0 power of zero, or 1, is to be subtracted; the next − to the left means, that the 1st power of 3 is also to be subtracted; the next, that the 2nd power of 3 is to be subtracted, and the + in the 4th place of figures means, that the 4th power of 3 is to be added. When any power belonging to any rank or place is not required in expressing any particular number, 0 occupies the place of either + or − in that place of the combination of characters for the number: thus + 0 − 0 − + would express 243 + 1, diminished by 3 and 27, or 214, to express which, the 3rd power of three, or 9, is not required; the 0 then in the third place expresses this, and yet gives the proper value to the −, 0, and +, in the 4th, 5th and 6th places. Arithmetical operations are performed by the aid of these simple marks with all the rapidity and security of the simplest algebraic processes, and pretty much in accordance with the well-known algebraic rules: thus to add 214 and 14 the process would be thus:

$$
\begin{array}{lll}
214 & \text{or} & +\,0\,-\,0\,-\,+ \\
\underline{14} & \text{or} & \underline{+\,-\,-\,-} \\
\text{sum } 228 & \text{or} & +\,0\,-\,+\,+\,0
\end{array}
$$

Here the + and − in the place of the 0 power of 3 destroy one another; if the two −'s in the next place had a third with them, they would go on as one − to the 3rd place; that − is then supposed to be introduced, but to balance it a + is also introduced and set down below: we then have two −'s in the third place, which similarly give + below, and one − goes on to the fourth place, where the + and − already existing balance one another, or mutually destroy the −; that which has been brought on appears below, and so do the 0 of the 5th place, and the + of the 6th , there being nothing to alter them, One other simple example must suffice: multiplication of 214 by 14 will stand thus.

$$
\begin{array}{r}
+\,0\,-\,0\,-\,+ \\
+\,-\,-\,- \\
\hline
-\,0\,+\,0\,+\,- \\
-\,0\,+\,0\,+\,- \\
-\,0\,+\,0\,+\,- \\
+\,0\,-\,0\,-\,+ \\
\hline
0\,+\,+\,0\,+\,0\,0\,0\,-
\end{array}
$$

Or 2187 divided by 81 minus 1 + 2996

 The translation of a given number into the ternary combination of signs suited to express it, requires the aid of voluminous tables, which may perhaps be simplified.

 In the machine, the successive rods have the power of the number that would be expressed by a +, or −, or 0, in the place in which the brass nail appears. The range of the machine is greatly extended by the use of the lines parallel to the zero line, for thus 3 +'s or 4 −'s may be in succession placed by the rod being carried down 3 lines or up 3 lines, and thus additions and subtractions are performed by the very motions which work the rods in the processes of multiplication and division. The extent of power of the machine will be conceived if we consider that 7 +'s followed by 480 can be set on the machine, and would represent the number

87104955839944780770790212;

while 55 +'s would be, if all the rods were set one line only below the zero,

872246055045600089535585253, or

87 quadrillions, 224605 trillions, 504560 billions, 89535 millions, 585253 thousands.

Appendix 11

Thomas Fowler/Francis Baily correspondence

By permission of the Science and Technology Facilities Council and the Syndics of Cambridge University Library.
Ref: Cambridge University Library. Royal Greenwich Observatory Archive. RGO60/4 Misc. Correspondence 1828-1843

Sir, I most respectfully beg to address a few lines to you in reference to the Calculating machine I had the honor to submit to your inspection in May 1840. Since that time I have made great improvements with it for all purposes of calculation, by constructing proper and convenient Tables for converting the results into the common denary notation, I find it now practically useful in a very high degree for all, even the most extensive calculations for the public lands, Interest, Commission, Ready Reckoner to the millionth part of one farthing – accurate division for Poor Law Unions etc etc etc. and even for the calculation of Logarithms or for finding the natural number to any Logarithm, which the Machine I exhibited in London is capable of in a singularly beautiful and concise manner to 12 or 13 places – finding however that the trouble of reconversion in to the decimal notation operated to the disadvantage of this mode of calculation I have given some attention to the other Lower scales, particularly to the denary and find that the principle of the same Machine can be extended to all scales of notation indefinitely, the only limit being the extent of human capability for the mechanical construction, – I have often for many years past contemplated on the advantages that would result in most cases of decimal fractions (or rather in fractions of the decimal form) if the radix of the scale were a multiple of the three first primes namely of $2 \times 3 \times 5 = 30$, and I have now no doubt whatever but that a Machine to the extent of this Radix may be constructed within a moderate compass that should involve at least 40 or 50 places of figures, or possibly as far as 100 places by means of accurate workmanship such as might easily be procured in London, – I have the whole construction of such a Machine in my mind's eye and could give a rigid demonstration of its practicability before a file or hammer be used, − this Machine would include and might be used for any Scale from Radix 30 down to the Binary and all the machinery of those scales would in a manner be in operation in a calculation by this splendid Machine, if the Scale of 30 were in operation − I have invented a very simple notation for the digits of this scale which consists of the 10 characters of common arithmetic and a few letters of the Alphabet under certain modifications and the whole is complete as far as concerns the notation, which notation is equally eligible for every other scale downwards − but my chief object at present being confined to the common decimal scale, I have made a model in wood of one term suiting this scale just as I should employ it in the Machine and I find its action perfect, & I am therefore certain that the principles of the Machine you have already so kindly noticed, can be fully adapted to the denary notation at once, and that by it, all Arithmetical operations can be performed mechanically with great despatch and certainty and the results instantly be presented if required as they are produced and stand in the Machine, perhaps I may construct a Machine to this extent with my own hands in the course of a few Months if God be pleased to spare my life and I have convenient opportunity, but my chief object in now writing to you is to beg that you will protect my claim to this invention. I am induced to do this more particularly in regard to yourself because when I had the honor to see you in London I felt the kindness and urbanity of your manner towards me very deeply. The same remark applied to Dr Buckland whom I have lately seen again on my return from the meeting of the British Association at Plymouth. The Dr with much feeling informed me of your late accident and narrow escape and gave me great pleasure by saying that you were now quite recovered from the effects of it − I have for a long time intended writing to thank you for presenting to the Royal Society some papers of mine transmitted to A. De Morgan by Sir Trevor Wheler and was rather surprised to find a note from this gentleman to Sir Trevor saying that he had received a pro-forma notice from the Council (or something to this effect) that no part of them would be published under its sanction, this was indeed nipping things in the bud as the papers were open to any alteration or improvement, but I had no desire for this honor, until the Marquess of Northampton, in your presence, advised me to draw up the best description of the Machine in my power on my return to Torrington and send it up, which I did and

I believe it was accompanied with a very lucid and able report by A De Morgan, Esq. which I have since seen and much approve − I have no doubt but the Council thought the whole unworthy of notice as the production of an humble individual like myself, but I may be allowed in extenuation of my temerity, to remark, that however humble a first Idea may be it often leads to the most valuable and refined results, and the very circumstance of giving it honorable publicity is the means of such further improvements as could not a first be anticipated, conclusions such as these may be drawn from the whole history of science, which is only a series of successive improvements on simple and original ideas. −

I should not have yet written to you in all probability had I not seen in the Times Newspaper of the 13th Instant an account of a new calculating machine by Dr Roth of Paris this may be a valuable Instrument or only a Toy, at any rate I now feel that I ought to be no longer silent in regard to my own invention as I am fully confident of producing an Universal Calculating Machine, a desideratum which has been so much sought after for many years past and of which the Machine I exhibited in London last year is the very first rudiment and foundation −

I hope Sir that you may not consider this Letter an impertinent intrusion, I have long looked on you as the best friend I could apply to, whose station and ability were sufficient to bring this matter into proper notice, and if the whole or any part of what I have now written after proper correction, be published in the Times and other papers under the influence of your name, it will draw attention to the subject, and I am perfectly free and willing to do any thing that may be in consequence required should any circumstance of this kind be found desirable −

I have the honor to be, Sir
Your most respectful,
Humble servant
Thos Fowler
Gt Torrington, Devon Oct 19th 1841

To
Francis Baily Esq,
London

Appendix 12

Fowler, Thomas. Description of the table part of the new calculating machine invented by Thomas Fowler of Great Torrington, Devon, in 1842: and which is now open for public inspection in the museum of King's College, London. Great Torrington: Printed by M. Fowler, 1844

By permission of the Senate House Library University of London. Ref: Mathematical Tracts 1848 [DeM] L° (B.P. 39)

THE following description of the Calculating Machine invented by the late Mr. T. Fowler, of Great Torrington, was dictated by him to his eldest daughter, a few days before his death. It contains a full and particular statement of the principle of the Machine, and the method of working it, and is sufficient to enable any person of ordinary ability, after an hour or two's study, to work a common sum in Multiplication, Division, or the Rule of Three. In the hands of a Man of Science, the principle of the Machine may doubtless be widely extended and applied practically to all purposes of Calculation.

It is now open for public inspection in the Museum of King's College, London.

DESCRIPTION of the Table Part of the New Calculating Machine invented by Thomas Fowler

Draw two lines from left to right, parallel to each other, about a quarter of an inch asunder: shade the space between them lightly with a pen : this space is called the Zero line: on each side of the zero line draw parallel lines at the same equal distance as the distance between the lines in the zero space. These lines may be continued on both sides the zero line up to any extent, according to the size of the paper. Take a point at the extreme left of the zero line, and with a pair of compasses make equal divisions of the line from left to right about a quarter of an inch asunder. Through these points of division draw straight lines parallel to each other and perpendicular to the zero line. Shade lightly the space between the first and second line at the left: let the next division remain open: shade between the next two lines, and let the next division remain open, and so on to the end of the paper, so that there is alternately a shaded space and a white or open space, throughout the whole length of the paper. The shaded lines represent guides, and the white lines between them grooves, in which sliders carrying brass nails at their ends move to and from the operator, and are made to move forward or backward by the operation of the machine. The zero line will consequently present a continued series of shaded spaces alternately parts of the spaces and guides, and when the brass nails are all in the zero line, the machine is at zero — (i.e.) *nothing* is represented on the machine. Let the zero line be marked 0 throughout on the guides, and the open line on each side of it 1 — and the next on each side of No. 1, be marked 2, on the guides, throughout, and then in the same successive order, 3, 4, 5, 6, and up to the extent of the lines parallel to the zero line.

The brass nails being all at zero, represent a line of 0s. in arithmetic; but if one of the brass nails be moved one division, or into the first white space towards the operator, and the decimal point be supposed to be on the first guide to the right of it, this nail represents 1, or an unit: on the second space it will represent 2 : on the third space towards the operator it will represent 3 : on the fourth space it will represent 4 : on the fifth 5 : and so on, as those spaces are numbered from the zero line towards the operator. These numbers are all positive or real numbers. Going back again to the decimal point, if the brass nail is placed on the first division from the zero line on the other side from the operator, it represents *minus* 1, or the want of an unit : in the space No. 2, it represents minus 2 : in the space No. 3, it represents minus 3 : and so on, up to the extent of the spaces or parallel lines, as they are numbered from the operator. If the decimal point is supposed to rest on the next guide to the right, instead of 1, 2, 3, the places of the brass nail would represent 10, 20, 30 etc. both positive and negative, according as the brass nail is on this or that side of the zero line : and if the decimal point be supposed to rest on the next division to the right, the brass nail will represent 100, 200, 300, 400, etc. etc. both positive and negative — and thus up to any extent of number positive as well as negative; consequently, any number may be placed or represented on the machine, as we have the corresponding places for the brass nails on the positive side of the zero line, including the 0s in the zero line itself. Similarly any negative number may be represented on the negative or off side of the zero line, by placing brass nails at the corresponding digit places, and also including the 0s in the zero line.

If now, a decimal point be supposed to rest on the first guide to the left of a brass nail, this brass nail becomes the first decimal place of a number, whether positive or negative : Thus, if the brass nail be on the space No. 1, towards the operator, and the decimal point on the first guide to the left of it, this nail would represent $\frac{1}{10}$: if it be in the second space, it would represent $\frac{2}{10}$: and so on, as the spaces are marked towards the operator: and if the decimal point be supposed to be on the next division to the left, the brass nail would represent $\frac{1}{100}$,

$\frac{1}{200}$, $\frac{1}{300}$, which are all positive fractions. But in this case, when the brass nail has one or more guides between it and the guide bearing the decimal point, an 0 or 0s, must be placed to the left of the brass nail according to the number of guides between the brass nail and the decimal guide: and as this is always uniform, the denominator being 1, with 0, understood to the right of it, the whole denominator may be entirely left out, and a dot set in place represented by the decimal guide; this dot is called the decimal point, and similarly with the negative or off side, or negative fractions, or secundal, tertial etc. fractions, according to the scale of notation used. If we now employ the decimal notation for the machine, we need use only 0, and the first five figures, namely, 0, 1, 2, 3, 4, 5, as any number whatsoever can be represented on the machine with these six characters only, by using the digits positively and negatively. Thus, put the decimal point on any convenient guide in the machine, then 1, 2, 3, 4, 5, may be represented with the brass nail as before described: but if we want to represent the number 6, we must employ the negative side of the zero line, and use the two sliders next to the left of the decimal point : the furthest to the left must have its brass nail set in the unit's place on the positive side, and the next slider must have its brass nail placed on the fourth division on the negative side from the zero line, which will then represent minus 4, and the two figures placed in their proper order will be with their proper signs + 1 − 4 = 6, as required. The number 7 will have the second brass nail in number 3, on the negative side : the number 8 would have it in number 2 : and 9 would have it in number 1 : and 10 would have it in zero : consequently, if we want to represent 16, the brass nail furthest from the decimal point at the left, must be placed in number 2 on the positive side, and the brass nail on the left next to the decimal point must be placed on the fourth space on the minus side, and we shall thus have + 2 − 4 = 16 ; 17 would be + 2 − 3 = 17 ; 18 would be + 2 −2 ; 19 would be + 2 − 1 ; 20 would be + 2 ; with the next brass nail to the right on the zero line.

We shall in future, after fixing the place of the decimal point, call the digits as they stand in any number, the first, second, third, fourth, fifth, sixth, etc. etc. places of digits to the left or right of the decimal point, and the digits themselves would be expressed as they stand in their respective places in the spaces on the positive and negative side of the instrument ; (but the extreme digit or 0 to the left in any whole number or decimal fraction, will always be called the first term of that number ; the digit to the right of this will be called the second term ; the third digit the third term ; and so on to the last digit or 0 to the end of the number.) Thus 16 would be 2 in the second place positive or first term ; and 4, in the first place negative or second term ; 200 will be 2 in the third place positive ; 500 will be 5 in the fifth place positive ; all to the left of the decimal point : and fractions in the same way to the right of the decimal point. Thus, .1 will be in the first place to the right of the decimal point : .01 will be in the second place to the right of the decimal point : .005 will be in the third place to the right of the decimal point : and in the same way with the negative numbers, whether whole numbers or decimal fractions.

Let us now represent the number 5231, on the instrument. Here four sliders are required on the left side of the decimal point. Place the brass nail on the first slider on number 1 ; the second slider on number 3 ; the third slider on number 2 ; and the next on number 5, in the fourth place to the left of the decimal point in the lines parallel to the zero line, or the positive side of the instrument. If the number 7321 be required to be placed on the instrument, as we shall here require the compound expression for 7, we must take five sliders to the left of the decimal point. We have already seen that the number 7 is composed of + 1 − 3 ; therefore the fifth slider on the left of the decimal point must be placed at + + 1, and the next at − 3, and the next at + 3, and the next at + 2, and the next at + 1, which will represent accurately the number 7, in this mode of calculation ; by which it may be seen, that when any digit exceeds 5, we must have the negative digit or digits in the same part of the number : and as in the denary scale every digit increases in value tenfold in succession from right to left, so ten divisions of any groove containing a slider will be equal to one division on the contrary way to the left or next place above it. This is like making carries by tens as in common arithmetic, provided the denary scale be in operation : nine divisions will make a carry in the ninary scale : eight in the octonary : and so on down to the binary or lowest scale that can be used ; and it matters not if these divisions be positive or negative, or compounded of both, provided the divisions are taken in succession, and the equivalent digit be taken in the contrary direction : this holds good whether the direction of the digits, and their equivalent carry, be positive or negative. Under this consideration, we may always adapt any number for operation in the instrument by the use of the instrument itself. Thus, suppose it to be wished to put the number 70853629 in the machine, in a way adapted for calculation; place this number altogether in the machine, proceeding by the first, second, third, etc. etc. places to the left on the positive side from the decimal point : now number 9 requires a carry, and carrying the brass nail backwards from number 9, ten divisions, the nail will rest in division number 1 in the negative side, consequently the first place of the number becomes minus 1, (which digit is distinguished from a positive digit by an oblique stroke of the pen through it, thus Ɩ) Now, having made a carry of 10 towards the negative side, we must bring its equivalent one division more towards or in the positive side, and thus the figure 2 becomes 3 : 16 again requiring a carry, we carry a brass nail ten divisions towards the negative side, where it will rest minus 4 : this 4 is distinguished likewise by an oblique stroke, and the carry of 1 added to the digit in the next place to the left makes 3 into 4, on which division the brass nail will now stand. The next term to the left is 5, which remains as it is. The next term to the left is 8, which requires a carry : carry the brass nail ten divisions to the negative side of the instrument, it will then rest on number 2 minus, to be represented as before with a stroke through it, thus ꝩ: the carry being again made, the 0 to the left is increased by unity, and the brass nail must stand at number 1, on the positive side of the zero line : the next term to the left, being 7, also requires a carry, and if the brass nail be carried ten divisions backwards from 7, it will rest on −3, (to be distinguished as before with a stroke 3ꝩ), and the next 0 to the left of it in the zero line must be brought to 1, in the positive side, for the equivalent carry. The given number is now completed, containing no digit in it above 5, and it will stand thus, 1 ꝩ 1 ꝩ 5 4 ꝯ 3 ꝩ, which contains one term more than the original number, on account of the carry being required in its first term.

The sliding rods have all a brass nail or counter at one end, and wires having a ladder-like appearance at the other : they are all perfectly similar to each other both as regards the distances of the wires from each other, and of the brass nail from the nearest wire. These distances must be most correctly measured, and the rods are required to move accurately in the plane of the instrument. The distances between the wires are given as will be seen by what I have called the radius of the instrument, and the angular motion of the moving frame in direction perpendicular to the plane of the instrument — by means of which, the rods are moved over 1, 2, 3, 4, or 5, divisions of the grooves in which the brass nails move, according to the magnitude of the respective digits in the number in operation. This motion is given to the sliding rods by small iron teeth, or flat wires, that rise always at first in a straight line between the wires, as they stand in any position of the rods.

The moving or lever frame is suspended on two central points lying in a plane parallel to the plane of the instrument, and enter the two sides of the lever frame about the middle, so that when the upper part is brought forward towards the operator, the lower part moves the same distance backward or from the operator, and vicé versá. The motion of the *upper* part of the lever frame *towards* the operator, is called the positive motion ; consequently the *backward* motion of the lever frame is called negative ; and contrariwise, if the upper part of the lever frame be made to move in a direction *from* the operator, its motion will be negative: the lower part consequently moving *towards* the operator, will have a positive motion. Thus, it appears that positive or negative motion can be given to the sliding rods according as the conditions of any number required, when the teeth fixed in these lever are brought up between the wires of the rods.

The machine I now have in mind is of four feet radius, which is probably much greater than needful, were it constructed of brass and iron : but the continual change in wood, by dryness and moisture, made me fearful of constructing it of a less size. The lever frame has 30 degrees angular motion about the centre, 15 degrees of which, are for taking the sliding rods forward or backward, in all cases. This motion is governed by proper stops, and is always of the same quantity — (viz.) 15 degrees either forward or backward.

I will now describe one of the levers as it stands in the lever frame, and its operation on a sliding rod above it, shewing how the motion of the brass nail is given for any digit in a calculation. The lever frame has the arc of a circle drawn on it, the radius of which is four feet, being the radius of the instrument. The middle point of this arc is in a straight line with the centre of the moving frame ; consequently a screw inserted into a hole at this point, will only revolve on its own axis, by the motion of the lever frame, and therefore cannot in any way affect the rods on the table part of the machine : This point is called the zero place of the lever frame. Let now about 15 degrees on each side of the zero point (making altogether 30 degrees of this arc) be divided with a pair of compasses into eleven equal parts, having the zero point most accurately in the centre of the middle division. The first points on each side of the zero point need not be noticed : the next point *above* it in the arc, is the place for the positive unit, and similarly the point at the same distance *below* the zero point, is the place of the negative unit : the next point above and below, are the places of the digits 2, positive and negative : and so on to the last or extreme points which are the places of the digit 5, both positive and negative. If small screws be inserted at these points, they will operate as fixed points to draw the lever backward or forward, according as the respective digits are required in any calculation. The five points then *above* the zero point are for *positive* digits, and the five points *below* it are for *negative* digits. Now, from the nature of this construction, if a bar move perpendicular to the plane of the lever in the direction of the zero point, and it be connected with the lever by a rod exactly four feet between the points of connexion, this lever rod will be the radius of the arc on the lever, when it stands at 15 degrees, or in the middle of its angular motion, and therefore if it be hooked on either of the screws, it will not move the bar either backward or forward : but if it be hooked on number 1, and the full motion of 15 degrees be given to the moving frame, this bar, carrying another bar perpendicularly upon it, into the end of which one of the before mentioned teeth is fixed, will, when the tooth is brought up between any two wires of a sliding rod, cause the brass nail to move one division or digit forward or backward : if the hook be fixed on either screw at number 2, on the lever, and the while motion of 15 degrees be given, it will cause the brass nail to move over two divisions ; and so on up to the screw at number 5, on which if the screw be fixed, the brass nail will move five divisions or digits in its groove between the guides in the upper part of the instrument. Thus by successive motions of the lever frame, any calculation is made by turning the winch connected with the sliding frame, that carries the moving frame from left to right, as the calculation proceeds.

By comparing what has been written with the machine itself, these motions and the construction of the machine will become apparent.

It remains now to shew, what a calculating machine of this kind, is capable of producing. The machine I have, contains about 15 or 16 sliding rods, and about 12 levers studded with the screws and fixed in the lever frame, which is complete as far as it goes, from left to right. A continuation of sliding rods and guides to the extreme right, on the upper board, would complete the upper part of the machine ; and levers with corresponding bars of wood in the lever frame, would complete the lever frame, if executed in precisely a similar manner to what already appears. Thus, the whole machine, to the extent I had intended it should reach, would be complete. The wood-work for this purpose, is already prepared, but fearing my health may not permit me to finish the whole, as I had first intended, I now feel compelled to anticipate the description ; but the completing it is as I have said, simply an accurate continuation of what has already been done. The machine, when fully complete will contain 45 places of figures in the upper part, and 27 places of figures in the moving frame or operating part of the machine, by means of which, any two numbers containing 18 and 23 places of figures, can be multiplied together. Division, also, to a great extent, can be performed with the same machine ; and even two or three operations of some extent in the Rule of Three, can be performed simultaneously, in a very curious and extraordinary manner. The sliding rods should be numbered 1, 2, 3, etc in succession, up to 47, from left to right : the grooves in which they slide may have the same numbers, so that if the sliding rods be taken out of the machine, they can be put back into their proper places. The levers in the lever frame must be marked 1, 2, 3, 4, and up to 27 from left to right, beginning with the inside of the frame to the left, which there serves as the first lever. Now, when the lever frame is brought up to the left as far as it will go, the numbers on the sliding rods and screw-studded levers will correspond, and this is called the first or commencing position of the instrument. Now set all the brass nails in their respective zero places : hook on all lever rods on the zero screws of the respective levers : the iron teeth at the extremity of the moving frame will stand nearly in a straight line, and if the iron compassing the moving frame be pressed downwards, the 27 teeth will enter the straight row of spaces between the wires above them. If now the moving frame be made to move on its centre, forward or backward, the whole of the sliding rods will still remain at rest, and consequently the whole machine stands at zero, ready to commence an operation.

Let it now be required to multiply two small numbers together — say 5 by 3: —

In all questions of multiplication, it may be better to consider every operation as an operation of proportion, or the Rule of Three; making unity one of terms. Thus, in the present case we may say, as 1 : 3 :: 5 answer ; and for the machine the three given terms may always be placed thus, ? | 3
5 | 1

which means as above 1 : 3 :: 5 ? the value of which note of interrogation will be found by the operation of the machine. The figures below the line belong to the moving frame, and the figures above the line to the upper frame. Now unhook any convenient lever to the right from its zero stud, and hook it on the first stud below : this rod will now be fixed on a negative unit in the moving frame, and serve as the first term or division in the Rule of Three : the 3 above it is represented by bringing a corresponding brass nail three divisions towards the operator, (which represents the factor 3 in the question) Go back to the left to the second studded lever : unhook its lever rod from zero, and hook it on the stud number 5, in the upper or positive part of the moving frame : this will represent the factor 5, in the given question. Now press down the lever frame with the left hand, by its iron handle, as far as it will go : this will set the moving frame at liberty to move its course of 15 degrees about its centre of suspension. Draw the upper part towards yourself, for the positive motion, and you will observe the brass nail at digit 3, move from you to digit 2 : at the same instant, the brass nail at zero, number 2, will move five divisions positively, or towards yourself. Now let the further part of the moving frame drop, and push the upper part from you until it locks itself at 15° on either side of its extent of motion. Again press down the moving frame in the same way with the left hand, and by the positive motion of the moving frame, reduce the brass nail at digit 2, to digit 1. This will cause the brass nail at 5, at the same instant, to move up to number 10, in the second groove. Repeat the operation again, and the digit number 1, will move into the zero line, and the brass nail will move from number 10 to number 15 : this requiring a carry, place the brass nail ten divisions towards the negative side of the instrument, and it will rest at the place of the digit number 5, and the next zero to the left being brought forward one division positively, as an equivalent for the ten divisions backward — the first digit is thus represented by the first brass nail, and is 1, and the second digit is similarly represented by 5, which together make the number 15, the answer or fourth term of the proportion, (viz.) 3 x 5 =15. This is one of the most simple operations of the machine. Suppose it be now required to multiply 5 by 31 : these factors are placed in the following order   ? | 31

5 | 1

: unhook as before any convenient lever from its zero stud, and hook it to the stud number 1 below, as before : hook also the second lever rod on its stud number 5, in the upper or positive past of the lever frame. Now set the brass nail corresponding with the lever bearing the negative unit, at digit number 3, towards yourself; set also the next brass nail to the right at number 1, towards yourself, and the machine is prepared for action. Repeat the same operation described above until you bring the digit number 3, to zero. Make the carry of 10 backward and 1 forward, as before. Give the winch one full turn, and the moving frame will thus be brought one term from left to right : press down as before, and pull the upper part of the moving frame towards you, and the digit 1, will be reduced to zero ; and the brass nail in the third rod, will, at the same instant, be brought up to the place of digit number 5, positive, and the answer will consequently be represented by 155, which is the product of 5 by 31 ; and similarly all operations in multiplication are performed — the only care requisite being to make the carries when needful. (This is always effected by an instrument over the sliding rods, made for that purpose)

NOTE. — In all cases and in every case that can be proposed, for the machine, it is absolutely necessary that the first term, whether unity or any other number, should reduce the number above it to zero or zero's, as the winch moves the frame step by step from left to right, and if the numerator and denominator of this fraction (which always stands to the extreme right) should be incommensurable, the answer or result for the numbers at the left, which are always found in the upper frame, may be continued in decimal fractions, should there be room sufficient for them : at any rate, if accuracy of fractions be required, a little ingenuity will shew that the remainders can be carried back again to the left, after writing off the results already produced. It is plain also, that if the denominator of the fraction be larger than unity, and both the other numbers greater than unity, the operation is one in the Rule of Three. Also, if either of the other numbers, not the denominator of the fraction to the extreme right, be an unit, the operation is one in division. A little practice by an intelligent person, will make this plain and beautiful, and such questions even as ?   ?   6874
784 632 5912

can be performed, if the moving frame can be made long enough to leave zero terms between the given terms, to take up the answers ; and the whole in a very curious manner is performed at once, by reducing the terms of the last numerator (which will be in the upper part of the machine) successively to zero by the action of its denominator, which will be in the moving or lever frame of the machine. The operation performed will be, as 5912 : 6874 : : 632 : ? : : 784 : ?, and the answers required for the two notes of interrogation are produced at once by the machine --the only care required being to look after the carries.

Also, if the denominator of the fraction to the extreme right be more or less than unity, and the numerator of this fraction and denominator of the fraction to the left, having the note of interrogation over it, be each of them an unit, the operation by the machine would give the reciprocal of the first term of the proportion or acting denominator, which reciprocal would be the answer required by the note of interrogation*.

* Thus far dictated to me, Caroline, E. Fowler, on the 25th. & 26th. days of February 1843, by my Father, when in great agony of suffering

It is necessary here to remark, that when the first digit at the left at the commencement, or as we proceed from left to right, is equal to or greater than the first figure of the divisor in the lever frame — that when the divisor frame is moved by the winch until the two first digits are brought together — that the dividend figure may be brought into the zero line : at any rate, the next move or two may be made generally to reduce the dividend digit an unit or to nearer to the zero ; and if it cannot be brought into the zero line, if the second digit from left to right on the upper frame be 8 or 10, and the digit that we want to bring into or nearer the zero line stands on the other side of the zero from the second digit, that a carry of 10 one way will cause a carry of the next superior digit one step nearer or into the zero line : and it will always be found that when a digit in the dividend is less than the first digit in the divisor, that the dividend digit can only be brought into the zero line by carries, in this manner, (viz.) the lever frame must stand one division more to the right, and its corresponding place with the smaller digit in the upper frame. A little experience will quickly shew all this, and the operation of division will be found exceedingly easy by

207

use of the carrying machine. It is precisely the same as when the first figure of the divisor is greater than the first figure in the dividend in common arithmetic. Thus, 584 divided by 94, has only one figure 6, in the quotient ; whereas, had there been a less figure than 5, for the first figure of the divisor, the quotient would have contained two places of figures in the whole number.

[Fowler.   Printer.   Torrington]

Appendix 13

The Reconstruction of Fowler's First Machine

Mark Glusker

Fowler's ternary calculating machine — an overview

Thomas Fowler, fearful that his ideas would be stolen by others, designed and built his wooden calculating machine entirely on his own in the workshop behind his printing business. To compensate for the limited precision that he could achieve in wood, the machine was very large: six feet wide by three feet deep and one foot high (180 x 90 x 30 cm). The model Fowler built has not survived, and almost all other evidence of the machine has disappeared. The descriptions of the machine that follow are based on our interpretation of how Fowler's machine might have worked, based on the limited information available.

Fowler had developed a technique that used balanced ternary to simplify complex monetary calculations for the Poor Law Union, publishing his methods in his book, Tables for Facilitating Arithmetical Calculations. (See appendix 2) His calculating machine was built several years later, giving mechanical form to the techniques outlined in the book. The choice of balanced ternary allowed the mechanisms to be very simple. Of course, it also required all the numerical values to be converted to balanced ternary, and then converted back to decimal at the end of the calculation. Clearly, this machine was not practical for simple addition or subtraction problems. Where the machine became really useful was for complex problems that required a great number of intermediate calculations in between the conversions to and from ternary. The calculations that Fowler faced at the Poor Law Union were exactly that sort of problem. (See an example of calculating with Fowler's Ternary Tables in Appendix 2)

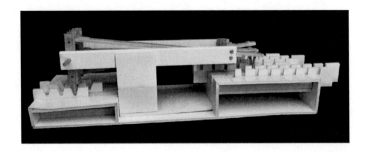

The first working model of Fowler's machine in over 150 years was the crude cardboard prototype that I made in late 1999, shown above. Although it was extremely fragile and finicky, it was able to correctly calculate a few simple multiplication problems in balanced ternary. This was a sufficient demonstration to warrant the development of a more precise model.

The Second Prototype

Like many calculating machines, the mechanism involves repeating the same components once for each digit of the machine. Due to the size of the machine, and the fact that it was originally made of wood, I chose to design all the parts so multiples could be easily made using an inexpensive, commonly available material: flat sheets of 1/8 inch thick pressed fibreboard. I had access to computer-aided design software (Pro/Engineer 2000i®) and was able to design an assembly that could be sent to a computer-aided laser to easily fabricate as many parts as required. Fowler envisioned a machine with 55 digits (although the exact number of digits in his 1840 prototype is unknown). I chose to model just a portion of the machine as a proof-of-concept, limiting the capacity of the machine to three digits in the multiplier and multiplicand, and seven digits in the product.

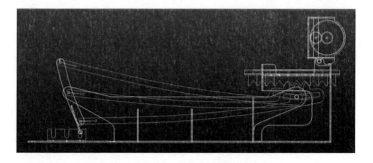

The prototype was designed on the computer and fabricated directly from the CAD database

In the context of a multiplication problem, Fowler's machine consists of four distinct sections:
− the multiplicand
− the multiplier
− the product
− the carry mechanism

The machine can also be used for division, but in reverse — the product becomes the dividend, the multiplier becomes the divisor, and the multiplicand becomes the quotient. For the purposes of the descriptions that follow, it is easiest to limit the discussion to problems of multiplication.

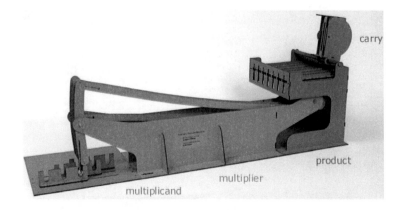

The Multiplicand

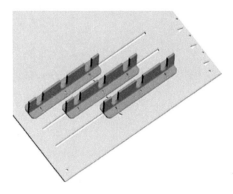

The multiplicand consists of a series of sliding rods, one for each digit of the multiplicand. Each rod can be set by the user to one of three positions, representing the three possible values for each digit of a balanced ternary number.

For multiplication problems, the rods are not really an active part of the overall mechanism. They are more of a mnemonic device to guide the user through the rest of the calculation. When used for division, the quotient is read on these rods.

The Multiplier

The multiplier is entered into the machine by clipping the end of each multiplier rod onto one of three rungs of a ladder-like rotating frame, one rod for each digit of the multiplier. These rungs represent the three possible values for each digit of a balanced ternary number: if a digit of the multiplier is a "+", the corresponding multiplier rod is clipped to the top rung of the rotating frame; likewise, for a "−" digit, the rod is clipped to the bottom rung, and a "0" digit is clipped to the central rung.

The rotating frame is mounted to an overall structure, using the central rung as the axis of rotation. The overall structure can slide laterally relative to the multiplicand and product rods, allowing the multiplier mechanism to act on each digit of the multiplicand one digit at a time.

At the opposite end of the multiplier from the rotating frame, the rods slide along a stationary cylindrical support. The rods each have a tooth that extends upwards to engage one of the product rods. As the ladder-like frame is rotated, it pushes or pulls the multiplier rods depending on the direction of rotation of the frame, and on which rung the rod is clipped.

The rotating frame has a tooth that extends to engage one of the multiplicand rods. The position of the multiplicand rod is an indication of which direction the rotating frame should be turned for that digit of the multiplicand.

The Product

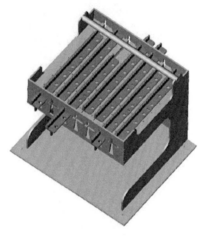

The product is a series of sliding rods, held on a raised platform at the other end of the machine from the multiplicand. They have teeth that extend below the platform to engage the corresponding teeth on the ends of the multiplier rods. As the ladder-like frame of the multiplier is rotated, pushing the multiplier rods one direction or the other, the product rods are pushed accordingly to represent the value of the result.

The initial position of the product rods is set to the middle of the platform, corresponding to a value of zero. The rotation of the ladder-like frame of the multiplier pushes or pulls any multiplier rods not set to zero. As a result, the teeth at the end of those multiplier rods engage the corresponding product rod, sliding it by one unit in either direction. As the calculation progresses, an individual product rod may be pushed more than one unit, causing an overflow in that digit. The platform supporting the product rods allows for an overflow of three units before the carry mechanism must be employed, bringing that digit back into the allowable range of "+", "0" or "−".

As the detailed design progressed, some surprising subtleties arose in what had appeared to be a very simple mechanism. For example, the tooth at the end of the multiplier rod had to be shaped carefully so that it pushed the product rod exactly one unit under a variety of conditions (see the diagram below).

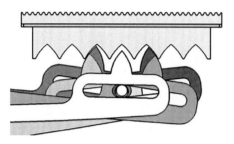

There were also some significant gaps in the available information. The carry mechanism, a feature of most calculating machines which allows an overflow in one digit to roll over to the adjacent one, was given only cursory mention by de Morgan. The carry mechanism developed for our model is perhaps the most speculative part of the design, as there was no evidence to suggest any particular approach to the problem

The Carry Mechanism

During a calculation, an individual product rod could potentially be pushed more than the one unit allowed for in balanced ternary notation. To correct this, a manually operated carry mechanism converts these overflows into the correct notation.

The carry mechanism works on only two adjacent product rods at a time. A gear train with a 3:1 reduction ratio engages racks on the tops of the product rods. As the lower order product rod slides by three units in one direction, the other higher order product rod slides one unit in the other direction. A three-unit long slot in the frame of the carry mechanism serves to guide the motion of the lower order product rod.

This carry operation would have to be performed several times during the course of a lengthy calculation.

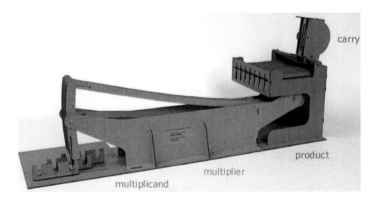

Instructions for use

These are the steps required to calculate a multiplication problem using Fowler's machine.

1)   Reset all the product rods to zero.
2)   Enter each digit of the multiplicand by sliding the individual multiplicand rods into position.
3)   Enter each digit of the multiplier by clipping the ends of the multiplier rods onto the appropriate rung of the ladder-like rotating frame.

4) Move the rotating frame on the multiplier to the vertical (neutral) position.

5) Slide the entire multiplier mechanism sideways to engage the tooth on the rotating frame of the multiplier into the slot of the first multiplicand rod.

6) Rotate the frame in the direction that would move the multiplicand rod to the zero position.

7) Slide the entire multiplier assembly sideways to an intermediate position, so that the tooth on the rotating frame disengages from the first multiplicand rod and ends up in the space between two multiplicand rods.

8) Repeat steps 4 through 7 for each digit of the multiplicand.

9) At the end of the calculation, or during the calculation if any of the product rods have an overflow of three units, engage the carry mechanism to eliminate any overflows.

Appendix 14

Fowler/Brougham correspondence

By permission of UCL Library Services, Special Collections.
Ref: Brougham Collection 46,624.

My Lord
As the time appointed for the commencement of the next Sessions of Parliament is fast approaching your Lordships public virtue and great talents will again be called into action, & I truly believe, for the benefit of the human race among other matters the Laws of Patents for inventions will in all probability be again taken into consideration and such amendments be made as may in future preserve the poor ingenious Man from almost certain ruin as the only reward of his ingenuity and as every species of information may be useful at the present moment I humbly and respectfully beg to lay before your lordship my case as a Patentee, a case which indeed sends me back angry & disappointed to an "almost starving Family".

In April 1828 I entered a Caveat for an invention for circulating hot fluids for various purposes by means of a bent tube which I called a Thermosiphon and soon after I applied for a Patent in the usual way, which after considerable delay in consequence of his late Majesty's health, passed the Great Seal 2 Oct. 1828. immediately I had notice from the solicitors Mess. Clowes &C to prepare the specification. I accordingly described everything relating to it honestly and fairly to the best of my knowledge, and very much desired to send my Manuscript to your Lordship, then Henry Brougham Esq, as Counsel, to prepare from it the specification; as I was then as now truly devoted to the honesty & nobleness of your Lordships public Character. I was however overruled and Mr. B. Rotch of Furnival's Inn was selected as my attorney, Mr. Smith of Torrington, informed me that those things were not in your Lordships practice, – Mr. Rotch summoned me to London towards the end of November as he could not prepare the specification without my assistance he then made an abstract from my manuscript (which is since printed & sold by Longmans & Co) with some little variation which was taken to Clowes & Co to be engrossed,

218

on further consideration I was not satisfied with his description as I believed it was not sufficiently explicit and therefore the next morning I went to Clowes & Co's office to make some addition but found the specification was engrossed and ready for my signature. I insisted on some further addition being made, leaving Mr. Rotch's language entire, and after some opposition I was indulged by adding another skin to the part already engrossed. this I was informed seemed to offend Mr. Rotch but I was determined to act according to my judgment, honestly & fairly by describing what I did then & do now consider a most useful and extensive application of the invention which I am certain will hereafter give much employment and prove highly beneficial when the opposition to this as to many other Patents by interested Men, shall be ended by the expiration of the short terms for which those monopolies are granted. I delivered the specification thus amended into the Chancery Office 29 Nov 1828 and returned home thinking all was safe, but soon after I was informed that Cottam & Hallam and others in London & other places were infringing my Patent openly & without reserve and having almost ruined myself with the expense & by experiments I could not "go to Law" to establish my right. I therefore wrote to Mr. Rotch, in forma pauperis, for his advice and was soon informed by this worthy, that the Patent was utterly vitiated by my addition to his specification and that it could not be supported in any Court of Law; & consequently that I had not other resource but to hang or drown myself according to any particular inclination I might have for either of those respectable modes of getting rid of all my troubles, I believe my Lord that thoughts like these are too often induced, and with fatal termination by the present Patent Laws; in my, perhaps not singular case I am sacrificed for being too honest but the most annoying circumstance is, that I am compelled to look on Mr. Rotch as a principal Agent in this affair as I am credibly informed that Cottam & Hallam applied to him for an opinion and that he told them they might infringe my patent with impunity, I do not even now believe that my specification vitiates the Patent but I have no means of bringing the matter before your Lordship in a legal form & must therefore content myself with the loss although poverty is absolutely staring me in the face in consequence of the Patent Laws and of the power the English Lawyers now have of making the worse appear the better reason; this Patent with expenses has cost me above £400 most of which I was obliged to borrow so that I may say I am ruined.

I humbly hope, my Lord, that what I have now written may be kindly received. I consider myself much injured by the imperfection of the Patent Laws, and I believe it would be better to repeal them entirely than to leave them any longer in their present state, the Pest of human ingenuity; I know my invention will be highly beneficial to my fellow Men and hereafter give constant employment to many. I have myself for hundreds of hours in solitude & silence during the last four years witnessed its operation and considered the many useful modifications of which it is susceptible and the more I have studied it the more I am convinced of the truth of my assertion and it is very hard that any Poor Man that ventures to take out a Patent for what he considers an useful invention and spends his all in obtaining it should be compelled to give up all the benefit originally intended by the legislature; or; immediately after become the Prey of the Caste of needy and too often rapacious Lawyers; at any rate it were better that the Government should have the Power to compound with the Patentee if his invention be useful and it be not proper that he should enjoy the monopoly for the time granted; as in this case something like justice would be done and the Patentee might be satisfied— —

I am My Lord
with the utmost respect
Your Lordships devoted
and humble servant
Thos Fowler.

Great Torrington Devon January 22nd 1834

Appendix 15

Fowler family tree through Hugh Fowler's line.

Courtesy of The Fowler family.

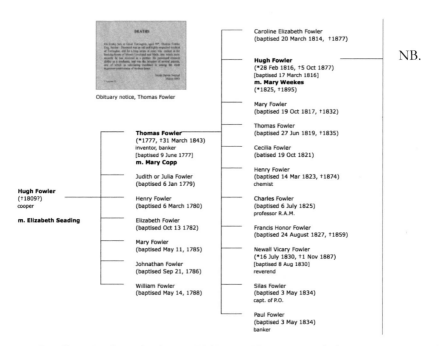

Caroline Fowler's death was 1853, not 1877 as stated above.

Sources

(i)   Archival Sources
(ii) Books
(iii) Other printed material
(iv) Web pages

(i) Archival Sources

British Museum
A.D. 1828 Patent No 5711 – Apparatus for Raising and Circulating
Hot Water Etc

British Association for the Advancement of Science
Transactions of the BAAS from 1831–1860. Transactions of the
Sections. Notices of abstracts.
   Vol 10 p 55 Mr. Fowler's New Calculating Machine
   Vol 11 pp 39–40 On a New Calculating machine. By Mr. Fowler
Journal of Sectional Proceedings No 1 11th meeting 1841 p2

British Library Manuscripts Department
MSS 371 Babbage Papers

University of Cambridge Library; Royal Greenwich Observatory
Archives. Material reproduced by permission of the Science and
Technology Facilities Council and the Syndics of Cambridge University
Library.
RGO 6 Papers of George Airy
427   48 Letter to Airy from Dr Buckland
         49 Letter from Airy to Prof. Phillips
         50 Letter from Airy to Prof. Forbes
         51 Letter from Prof. Phillips to Airy
         54 Letter from Fowler to Airy
         56 Letter from Airy to Sir Trevor Wheler
         58, 59, 63,65,67,68,69,71,72,73,74
             Papers re Babbage
440   118..124   Report re Analytical Engine
         385 – 568..571 Re Prof. Moseley

432 − 125   Scheutz calculating machine
428   J.W.Woolgar on Mechanical Computation
RGO 60 Papers of Francis Baily
60/4 Misc correspondence 1828−1843. Letter from T. Fowler

Devon Record Office
Census returns
1891, 1881, 1871, 1861, 1851, 1841

Buckland Collection
Sir Trevor Wheler Letters & Papers 1813− 1852
1148 M/8/17   Paper on N D registration
Letter from Sir Trevor Wheler re Fowler 31.12.1841

Clinton Papers
96M Box 9/1B   Volunteers mock battle 1805
96M Box 14/2   Hugh Fowler Tenant of Lord Rolle
96M Box 108/15 Deed between Clintons and Thomas Fowler (cooper)
               1855
96M 21−22   Catalogue Stevenstone Library
D929.2/ROLO   Rolle family of Stevenstone

Illustrations
LD P&D05632 Gendall. John Torrington bridge 1840
LD P&D05633 "   " Torrington church 1840
P&D085335   Dawson W. Aqueduct over the Torridge 1849
H/B/PA249   Payne W. Torrington reproduced in Payne's Devon
             Payne Vol2

Torrington Parish Papers
2558M/4/1A   Council Minute Book
2558M/4/3   Book of elections
2228M/4/8   Watch Committee Minute Book
2558M/5/2   Borough Treasurers Accounts
2558M/8/6   Market Book
2558M/11/35   High Street Engraving
Gen 15/1−4   Order to form Poor Law Union 1835−1836
929J/GRE/GRE0 Great Torrington Parish Registers

MF−17 21.2.1813 Marriage of T. Fowler & Mary Copp
4014A/PW2   Church pew allocations
2558/8/1A   Torrington town council record book

Torquay Parish Registers
AD/TOR 06 DYM   Chronological record of Torquay.

276C/EFM 32   Hugh Fowler appt. Headmaster Bideford Grammar
school
LD BBT/A1/6/8 Deeds of Bideford Grammar School
TD 156/A4   Receivers account book 1829−1850. Payments to
Fowler as organist, printer & ironmonger. His
signature.
4014/add/   Charities report/Bluecoat school/church organ/post
office
1142B/T44/2   Thomas Fowler Scrivener. 1811
Z10/41/14   Thomas Fowler letter re financial matter. 1815
1142/B/T43/11 Thomas Fowler of GT Stationer. 1820
TD 156/M3   Younger Thomas Fowler appointed church organist
BB54/97   Burial Board payments to Mary and Cecilia Fowler
HMB 79/95, TC Bk 3, TC Bk 2. Misc payments to Mary & Cecilia
Fowler
Q/S63 18822−26 Fowler Printed publicans recognisance's

Gloucestershire Archive
B243/7447GS   Memories of the College School, Gloucester,
Frederick Hannam-Clark
GAL/D4*B412/37536GS   The King's School, Gloucester.
David Robertson

Great Torrington Museum
Council Declaration Books
T. Fowler elected councillor   05.11.1835
04.11.1839
05.11 1842
Map of Great Torrington 1843
Portrait of John Lord Rolle

Institute of Civil Engineers Archives Service
Tracts 8vo
Vol 58   Fowler's Thermosiphon pamphlet
Vol 96   Rev H. Moseley MA ,FRS Canon. On a machine for
                calculating the Products, quotients, logarithms &
                powers of numbers

Proceedings of the Institute of Civil Engineers
Vol 1 (1837)   Ref Cottam on Warming and Ventilation
Vol 11 (1852)   Memoir of the Marquess of Northampton
Vol XV 91855)   Re Scheutz Difference Engine & Babbage's notation

King's College, London – King's admin records.
KA/C/M4   Council Minute
11/1   Deposition of Babbage's machine in the museum
19/1   Prince Albert opening the Museum
7/43   Deposition of Fowler's machine in the Museum
OLB 1   Deposition of Fowler's machine
OLB 3 1850-54   Letter to Hugh Fowler re removal of machine
K/MUS 20   Museum Visitors book
K/MUS 21   Museum Curators Diary
KA/IC/F17   Enquiry from Hugh Fowler
KA/IC/F22   Enquiry from Hugh Fowler
C.M. Book M. Wheatstone Collection – Box 5
                Reference to 8 reprints of T. Fowler's Calculating
                Machine. (Reprints not found)

Royal Society
Archive papers
AP.23.24 De Morgan, description of a calculating machine...
AP.23.26 Thomas Fowler on his calculating machine
        Journal Book of the Royal Society Vol. XLV111 1836–1843
        (JBO 48) pp 622–629

Proceedings Vol. 4 450. John Herschel one of the first to observe the
        great comet (17.03.1843)

Misc. correspondence

MC3 1840 – Background information on procedure

Lubbock Papers
Vol. 10 C479   Letter re Fowler's demonstration
Vol. 10 C493   Letter re Fowler's demonstration

University College Library
Brougham Collection 46,624   Letter from Thomas Fowler

Senate House Library, University of London
Fowler, Thomas. Description of the table part of the new calculating machine invented by Thomas Fowler of Great Torrington, Devon, in 1842: and which is now open for public inspection in the museum of King's College, London. Great Torrington: Printed by M. Fowler, 1844 University of London. Mathematical Tracts 1848 [DeM] L° (B.P. 39) (De M) L8(Scheutz) – Re Scheutz Difference Engine

Wren Library, Trinity College, Cambridge
Whewell Papers
Add. Ms.a.210 23–40   Northampton Letters 1832–1847
Add. Ms.a.202 98   De Morgan letter

(ii) Books

Alexander, J.J. and Hooper, W.R. *The History of Great Torrington in the County of Devon.* Norwich: Jarrold & Sons, 1948

Austin, Anne. *The History of the Clinton Family 1299—1999.* Devon: Lord Clinton,1999

Acland, Arthur H, ed. *Sir Thomas Acland — Memoir and Letters.* London: Private circulation.

Babbage, C. *Passages from the Life of a Philosopher.* London: Longman, 1864

Barnes, J. And A; S. Scrutton — *Great Torrington & District Through Time.* *2014.* Amberly Publishing.

Barrett, C.P. *The Overseers Guide and Assistant containing plain instructions to overseers of parishes in Poor Law Unions.* 1840

Bathe, Greville and Dorothy. *Jacob Perkins. His inventions, his time and his contemporaries.* Philadelphia: Historical Society of Philadelphia, 1943

Bryant, Arthur. The *Age of Elegance, 1812—1822.* London and Glasgow: Collins, 1950

Charity Commission. *Charities of Devon 1826—30* Exeter: Besley, 1851

Christie, Peter. *North Devon History.* Bideford, 1995

Colby, F.T. *Torrington Worthies.* Read at Great Torrington August 1899.

Cresswell, Beatrice F. *Some notes on the history of the Parish Church of St Michael Great Torrington.* Exeter, Devon. Southwoods, 1929

Dickens, Charles. *Oliver Twist.* London: Penguin 1994.

Dickens, Charles. *A Christmas Carol.* Temple Press, 1946

Dictionary of National Biography. Oxford: Oxford University Press

Doe, George M. *Some Great Torrington Sticks and Stones.* Bideford, Devon: Bideford & North Devon Gazette, 1928

Dredge, John I. *Devon Booksellers and Printers in the 17th and 18th Centuries.*

Ereira, Alan. *The People's England.* London: Routledge & Kegan Paul. 1981

Fisher, Arthur. *Blundell's Register — Pt 1 1770—1882.* Exeter: Commin, 1904

Fletcher, I. ed. In *the Service of the King.* Spellmount, 1997

Fowler, Hugh. *Biographical Notice of the late Mr Thomas Fowler of Torrington with some account of his inventions.* Rep Trans. Devon Ass. Advmt. Sci. 7. (1875): 171—78

Fowler, Hugh. *Auxilia Graeca.* Gloucester, 1856

Fowler, Thomas. *Tables for Facilitating Arithmetical Calculations intended for calculating the proportionate charges on the parishes in poor law unions.* Gt Torrington, Devon: Fowler and London: Longmans, 1838

Fowler, Thomas. *A Description of the Patent Thermosiphon with some modes of applying it to Horticultural and other useful and important purposes.* London: Longmans 1829

Freeman, Benson R.N. *The Yeomanry of Devon 1794—1927.* ed Earl Fortescue, K.C.B. London : St Catherine Press,1927

Gardiner, Juliet, and Wenborn, Neil ed. *The Companion to British History.* London: Collins & Brown 1995

Gilbart, James W. *The History and Principles of Banking.* London: 1834

Gill, Crispin. *Plymouth A New History.* Devon Books 1993

Hathaway, E. ed. *A Dorset Rifleman.* Swanage, Dorset: Shinglepicker, 1995

Hathaway, E. ed. *A Dorset Soldier.* Swanage, Dorset:.Shinglepicker, 1995

Hannam-Clark, Frederick. *College School Memories,* 1890

Hearnshaw, F.J.C. *The Centenary History of Kings College* London 1828—1928

Heulin, Gordon. *Kings College London 1828—1978.* A History Commemorating the 150th Anniversary of the Foundation of the College

Hibbert, Christopher. *The English. A Social History. 1066—1945.* London: HarperCollins, 1994

Hood, Charles. *A Practical treatise on Warming Buildings by Hot Water.* London: Whittaker & Co,1837

Hyman, Anthony. *Charles Babbage: Pioneer of the Computer.* Oxford: Oxford University Press, 1984

James, Frank. A.J.L. *The correspondence of Michael Faraday, Vol. 2.* London: Institution of Electrical Engineers, 1993

Johnson, D. Sketches of Field Sports as practised by the Natives of India. Gt. Torrington, Devon: Fowler, 1822

Johnson, D. *Observations of Colds, Fevers and Diseases of the Liver.* Gt. Torrington, Devon: Fowler, 1823

Johnson, Paul. *The Offshore Islanders.* London: Phoenix, 1995.

Knuth, Donald E. *Semi−numerical Algorithms The Art of Computer Programming Vol. 2*. Reading. Mass: Harlow Addison−Wesley, 1997

Klimenko, Stanislav V. *Computer Science in Russia: A Personal View*. In IEEE Annals of the History of Computing, Vol 21, No 3, 1999

Liddell Hart, B.H. ed. *The Letters of Private Wheeler*. London: Windrush, 1997

Massingham, Hugh & Pauline. *The London Anthology*. London: Phoenix, 1950

Maxted, Ian. *Devon Book Trades Bibliographical Dictionary*

Maxted, Ian. *In pursuit of Devon's History*. 1997

Moseley, Maboth. *Irascible Genius: A Life of Charles Babbage, Inventor*. London: Hutchinson, 1964

Palfreyman, J & Swade, D. The Dream Machine London: BBC, 1991 *Par A***.Pratique de l'art de chaufer par le Thermosiphon on calorifere a chaud...* Paris: Audot, 1844

Pearse Chope, R. ed. *Early Tours in Devon and Cornwall*. Newton Abbot: David & Charles, 1967

Quennell M. & C.H.B. *A History of Everyday things in England. 1733−1851*. London: Batsford Ltd. 1961

Ramsden, Charles. *Bookbinders of the UK outside London 1780− 1840*. London: Queen Anne Press, 1954

Rea, J. *Flora Ceres & Pomona*. Richard Marriott, 1663

Roberts, Cecil. *And so to Bath*. Stratford Press, 1940

Robertson, David. *The King's School, Gloucester*. London & Chichester: Phillimore, 1974

Schaffer, Simon. *Babbage's Dancer and the Impressarios of Mechanism*. In Cultural Babbage: Technology, Time and Invention, ed. Francis Spufford and Jenny Uglow. London. Faber & Faber, 1996.

Smiles, Samuel. *Lives of Boulton and Watt*. London: 1865

Snell, F.J., Blundell's. *A short history of a famous West Country School*. London: Hutchinson.

Sobel, Dava. *Longitude*. London: Fourth Estate, 1995

Swade, D. *Charles Babbage and his Calculating Engines*. London: Science Museum, 1991

Swade, D. *'It will not slice a pineapple' from Cultural Babbage, Technology, Time and Invention* ed Francis Spufford, Jenny Uglow. London: Faber & Faber, 1996

Swade, D. *The Cogwheel Brain*. London: Little, Brown & Co, 2000

Swete, Rev J. *A Tour of Devon*. 1789

Taylor, Henry. *Observations on the current coinage of Great Britain*. London: Groombridge & Sons. 1846

Timbs, J. *Curiosities of Science*. London: Crosby, Lockwood & Co., 1860

Timbs, J. *History of Clubs & Club Life*. Piccadilly, London: John Camden Hotten, 1872

Toole, Betty A. Ada, *The Enchantress of Numbers*. Marin County: Strawberry Press. 1992

Tredgold, Thomas. *The Principles of Warming and Ventilating Public Rooms*. London: M. Taylor, 1836

Tredgold, Thomas. *An Account of some experiments on the Expansion of Water by Heat*. in Transactions of Institute of Civil Engineers. Vol 1. XIV. 1836

Trevelyan, G.M. *British History in the Nineteenth Century*. London: Longmans. 1922

Trevelyan, G.M. *Illustrated Social History: 4*. London: Longmans, 1949.

Turck J.A.V. Origin *of Modern Calculating Machines*...Chicago: Western Society of Engineers, 1921

Venn, J.A. *Alumni Cantabrigienses Part II* 1752−1900

Vicary, B., *Art thou a Vicarie?* Devon

Walford, E. *Old and New London*. London, Paris & New York: Cassell, Petter & Galpin. 1880s

Ward, J. *Ward's Mathematicks*. London: Longman, 1758

(iii) Other printed material

© **The British Library** Ref Shelfmark RB 23.b.6538

The Athenaeum 1840−1844

1840 p 739   Account of papers at BAAS meeting

1840 p 845   Paper on cause of increase in colour. Sir David Brewster

1840 p 870   Mr Airy on Fowler's machine

1841 p 594   Notice of papers handled by BAAS

1841 pp 603−605 Opening of Plymouth BAAS AGM

1841 p 671   Prof. Moseley on a calculating machine

1841 p 672	'Mr Fowler's machine put up by mistake at the Naval Annuity Office'.
1841 p 700	Account of Fowler's demonstration of the machine at Plymouth
1841 p 715	Report on Perkins system of warming buildings by hot water
1841 p 734	WDC letter re solution to necessity for tables
1843 p 594	Re opening of George III Museum
1857	Scheutz letter re Difference Engine

Devonshire Association – Transactions

1839 Vol 7 p 100	Colby. Spire collapsed and damaged organ donated by Lord Rolle 1809
TDA 71 p61	Re Caroline Fowler
TDA. 71 p 47	Henry Fowler
TDA 1867 p187	Henry Fowler on opening of barrow at Huntshaw
TDA 1874 p370	Biographical notice, Henry Fowler
TDA 1875 p 171	Biographical Notice of Thomas Fowler
Vol. 10 1878 p55	Biographical note, Hugh Fowler
Vol. 11 1879	R.N.Worth. Notes on the History of Printing in Devon
TDA 1899 p174	Biographical notice, Daniel Johnson
Vol. 56 1924	Obituary Rev William Weekes Fowler
TDA 1990, 1–16	Presidential Address by F.J.M. laver 'Computers, Telecommunications and Information technology'.
TDA 1934 67/208 & 68/136	Re Miss Editha Pring Fowler
TDA 1992 p155–6	Hugh Fowler defending Torrington's glovers

Devon & Cornwall Notes & Queries

Vol. 1 48	Bull ring at Great Torrington
Vol. 16 14	John Milford of Coaver. Bank note at Torquay Museum
Vol. 21 1940–41	George M. Doe
21/56	Gt Torrington church sundial/memorial window/brass tablet to Henry Fowler

Devon Probate Office

Wills	Hugh Fowler	October 1877
	Charles Fowler	June 1891
	Cecilia Fowler	May 1883
	Henry Fowler	July 1874
	William Fowler	September 1881
	William Weekes Fowler	July 1923
	Editha Pring Fowler	June 1949
	Thomas Hugh Fowler	May 1970
	Mary Jane Fowler	December 1907
	Mary Elizabeth Stodden Fowler	Nov 1925
	Benjamin Knight	August 1923

© **The British Library** Shelfmark ST 1894
The Gardeners Magazine 1826–1843
Vol. 1

p 37	Tredgold on moisture & evaporation in artificial atmospheres
p 167	Atkinson on management of hothouse fireplaces

Vol. 3

Preface	Summary of developments in hot water heating
pp 186–192	Wm. Whale's installation for A. Bacon
pp 365–368	Methods of heating hot water for greenhouses
pp 423–432	Experiments in heating by hot water by William Atkinson

Vol. 4

pp 17–19	Cottam & Hallam – improvements in circulating hot water
pp 19–20	Heating of hot houses – Henry Bains
pp 28–31	Chabbannes pamphlet on heating by hot water
pp 61–64	Comment on above papers

Vol. 5

pp 20–23	Byers on heating hot houses by hot water
pp 260–263	Byers – further improvements
pp 453–454	Review of Fowler's system
pp 544–545	Mr Weekes – Heating by hot water

Vol. 6
pp 334–335   Henry Dalgleish on Fowler's Thermosiphon
pp 373–377   Joseph Knight on heating by hot water.
Vol. 7
pp 82–84   Reviews of various systems
pp 171–185   Thomas Tredgold on heating by hot water
p   612 Review – Cottam & Hallen, & Fowler
pp 685–686   Review Kewley's system
Vol. 8
pp 292–297   M. Perkins, Plan for heating by hot water

Gentlemen's Magazine
1835 3/6   Lord Brougham re Bill to amend Patents
1835 – Sept.   BAAS AGM in Dublin
1835 pt2 p556   Obituary, Daniel Johnson
1839 Oct   Charles Babbage resigns from BAAS
1840 pt2 p525TD Acland appointed VP for Plymouth meeting of
            BAAS
1842 p419   Sir T. Wheler to be Lt Col Royal N. Devon Yeomanry
1851 p543   Sir T. Wheler app. Commandant N. D. Yeomanry
1850/74   Hugh Fowler master of Bideford   Grammar School

Journal of the Plymouth Institution
1878   Re country fairs and Institutions

NATWEST Group Archives
10698, 10699   Rotation Committee Minute Book. Take–over of
            Loveband & Co by National Provincial

London Metropolitan Archives
MF72   London Directories
Johnstone's London Commercial Guide & Street Directory 1818
Robsons London Commercial Directory. 1828
Robsons London Directory 1842
Kelly's 1840
Pigots 1838

© **The British Library** Shelfmark RB 23.b.6538

The Mechanic's Magazine 1828–1845
Vol. 8
pp 226 – 229 System of heating by water: Illustrations &
Improvements
pp 337–341 New system of warming by hot water – Wm Whale.
pp 392–393, 415, 425–426 Comment on new system
Vol. 10
p 64   Re Babbage's machine
p 139   J.W. Woolgar, a companion to the common sliding rule
p 320   Notice of Fowler's Thermosiphon Patent
Vol. 12
Pp 368–384 Select Committee Report on Patents for inventions
Vol. 14
p 203,217,227 Review of 'Reflections on the Decline of science in
England'. (Continued in subsequent volumes)
Vol. 16
p389   Comparative advantages of heating by hot water, hot air &
steam
Vol. 20
P 393   Article to prove not pressure of the atmosphere alone which
raises water in a vacuum to the height of 33'
Vol. 32
p 223   Perkins Patent 16.12.1839
Vol. 34
p 346   J.W. Woolgar on a calculator
p 386   Re Perkins systems of heating by hot water
Vol. 35
p 142   Perkins patent 1841
p 336, 346–347 Re Dr Roth's calculating machine
p 510   Comment on Roth's machine and another
Vol. 36
p 446   Dexter Patent for heating by hot water

Newspapers
Devon & Exeter Gazette
31.12.1829   Re Thermosiphon

Exeter Flying Post

11.04.1805 1d	R. Wilson to retire.
23.07.1807 2c	R. Wilson selling business
28.01.1819 1c	T. Fowler lottery agent
18.01.1821 1d	T. Fowler lottery agent
25.12.1828 3c	Cross House for sale
08.01.1829 3b	T. Fowler – patent for hot liquid systems
03.09.1829 2d	T. Fowler obtained patent for Thermosiphon
10.09.1829 2e	Thermosiphon being built in Exeter
31.12.1829 2c	Report of Thermosiphon
25.02.1830 1c	Cross House to let
21.06.1832 2d	T. Fowler. Daughter died
23.10.1834 3a	T. Fowler appointed Inspector of Weights & Measures
09.06.1836 2d	T. Fowler. Son died
06.04.1843 2f	T. Fowler obituary
24.04.1907	History of Rolle's at Stevenstone

Exeter & Plymouth Gazette
30.12.1829/4.1.1830 Re Thermosiphon

Gloucester Chronicle
11.8.1877   Re Hugh Fowler

North Devon Journal
21.02.1839	Petition to repeal Corn Laws to Lord Brougham Notice of meeting to petition Lord Rolle and Lord Clinton in favour of the Corn Laws
31.08.1839	Cricket match on Northam Burrows
24.09.1840 3b	Quicksilver mine
09.1840	Lord Rolle 90
06.04.1843 3d	Deaths. T. Fowler
Sept 1853	Fowler on decimal coinage
15.06.1854	Hugh Fowler's farewell address on leaving Bideford Grammar
24.05.1855	Newell Vicary Fowler elected Fellow of Sidney Sussex College
17.03.1859	Re Hugh Fowler
03.07.1862	Restoration of Torrington church

27.01.1870   Mr Fowler, postmaster, having telegraph installed
08.1875   Hugh Fowler's address to the Devonshire
                    Association

Plymouth, Devonport & Stonehouse Herald Times
1840 – 13.5a, 20.4e, 22.6e. Re calculating machine
13.10.1941   Dr Roth & his calculating machine

Western Daily Mercury
28.07.1875   Re Devonshire Association AGM

Western Morning News
04.04.1936   Link established between Dick Whittington and Gt
                    Torrington

Western Times
19.02.1839   Petition to be sent to Lord Brougham on the corn
                    laws.
24.2.1856   Re Obelisk commemorating Napoleonic Wars

Woolmers Exeter & Plymouth Gazette
02.09.1840   August Bank Holiday Cricket.
09.09.1840   Yeomanry assemble for 8 days.
10.10.1840   Torrington Fair
25.03.1843   Re Comet

Trade Directories
Billings Directory – 1857
Harrods Directory – 1878
Kelly's Directory – 1923, 1919, 1914, 1910, 1906, 1902, 1893, 1883,
                 1876, 1873, 1866, 1830
Morris's Directory – 1870
PO Directory of Devonshire – 1856
Pigots Directory –1830, pre 1830
Slater's Directory
Torbay Directory
White's Devonshire Directory – 1878, 1850

Westcountry Studies Library
Cuttings file on Great Torrington
    Print of Stevenstone
    1834 report on Torrington
    Re Gaol, lighting, state of roads, public houses, church pews
    Criticism re poor management of Almshouses Trust
    Gloving flourishing, population statistics
    Doe, George,M. 'Some Bits of an old Borough'

Plymouth & West Devon Record Office
A 629   In 1829 Royal Annuitants Soc Actuary was a Lt Somerville.
        Kerr Street. Devonport

(iv) Internet Resources

http://www.mortati.com/glusker/
Mark Glusker's concept models of Fowler's ternary machine
http://www.thomasfowler.org.uk
Information on Thomas Fowler
http://www.thomasfowler.org.uk/calc.htm
Stained glass window representation
http://www.thomasfowler.org.uk/thermo.htm
The Thermosiphon
http://www.zyvex.com/nanotech/babbage.html
Links to:
Fowler – ref Dr Doron Swade MBE
Molecular mechanical computers
Babbage
http://www.computer-museum.ru/english/setun.htm
N.P. Brousentsov, S.P. Maslov, Alvarez J. Ramil and E.A.
Zhogolev, "Development of Ternary Computers at Moscow
State University," Russian Virtual Computer Museum
http://www.dcs.warwick.ac.uk/bshm/
The British Society for the History of Mathematics
http://www.cbi.umn.edu
Charles Babbage Institute
http://www.merkle.com
Homepage of Ralph Merkle
http://www.royalsoc.ac.uk/
The Royal Society
http://www.great-torrington.com
Great Torrington
http://blpc.bl.uk
British Library
http://www.ulrls.lon.ac.uk/resources/MS913.pdf
Augustus De Morgan archive

Reader Reviews for Seeds of Doubt

"I finished Seeds of Doubt last night. It was a great book… I loved the story, Ingrid's voyage and struggle, and the sense of Devon and history. The research struck me as particularly impressive…thanks for the entertainment of a great book."

Simon Hall — Home Affairs correspondent for BBC South West, *www.thetvdetective.com*

"Thanks for a Great Book, can't stop reading it."

"To build a conspiracy thriller on an event as well known as the Lynmouth flood disaster, and to make it both plausible and exciting, is no mean feat. The dialogue is well written and convincing and the plot is unpredictable with enough twists to keep any thriller reader happy."

"I've just finished Seeds of Doubt; WONDERFUL. I loved it; it has all the right ingredients — drama, tension, intrigue, love, secrets, danger and truth. Looking through the references, why hasn't this made more of a splash (unintended pun) in the national and local press? It's huge. Thank you so much for a thought provoking and thoroughly edge of the seat read."

"I couldn't put (it) down! It is an amazing story, vividly written … some of your insights and or descriptions are so 'word perfect' and then there are the subtle sub plots! The actual theoretical basis is very disturbing − terrifying? — and so it took some hours in the day to shake off from being inside the book. Remarkable. Brilliant! Thrilling. Enthralling…"

"Just had to let you know that I REALLY enjoyed your book — it is well written & constructed, and a very good read."

"The main character of this intriguing story is a woman who experienced at first hand the disastrous flood at Lynmouth in 1952. Some thirty years later she still has difficulty in coming to terms with the events of that fateful night, particularly when, in the course of her job as a provincial journalist, she comes across material which suggests that the disaster may not have been an act of God but man-made. Her urge to find out the truth brings her into conflict with her boss and her husband, and attracts unwelcome attention from government agencies. She is forced to confront issues from her earlier life, and finds her own life in danger again.

With many unexpected twists to the story, this is a book which I found difficult to put down. With an appendix of references to official documents, one is left wondering whether the Lynmouth flood was really a freak of nature or something more sinister."

"I thoroughly enjoyed this book. The way that fiction was blended with fact was very well done, and the list of references to the official documentation really makes you think. Living in the local area, there has always been speculation about the cause of the flood. Have already bought this for friends!"

Seeds of Doubt
Chapter One

March 1982, Salisbury, England

'Rainmaking!' Nick Pearce pushed the café door open and threw his dripping coat over the first available chair. 'Are you seriously saying that...?'

'Not me,' Ingrid interrupted. 'Private Dean.'

'Who's he?'

'The old boy in the photo.'

'Oh, yeah.'

Ingrid's heart sank. Her boss's off-hand response wasn't exactly encouraging. She edged round the marble bistro table to the chair opposite. 'I interviewed him yesterday for the British Legion piece. Here,' she leant down and retrieved the black and white photograph from her bag, 'there's a cutting stuck to the back all about rainmaking. And before you ask, it's nothing to do with voodoo or ancient pagan rituals. He swore he'd seen planes make it rain when he was on manoeuvres on Salisbury Plain.'

Nick slid the photograph towards him, flipped it over and scanned the text on the yellowing piece of paper. 'So according to this, planes flew overhead, the clouds got heavier and blacker and about thirty minutes later it rained.'

'Bit more to it than that, but yes.'

The silence lengthened. Taking a deep breath, Ingrid finally summoned up the courage to get to the point. 'What do you think of Private Dean's story?'

Nick was lost in his own thoughts. 'What?'

'Private Dean. What do you think?'

'I think it's an old boy winding you up.'

'He was pretty convincing,' she said.

'Yeah?' Nick muttered, swilling his free cappuccino around an over-sized cup.

'I'd like to follow it up.'

He looked up. 'Why?'

'Why not?'

'Because our readers want to know what happened last week, not thirty years ago.'

'I still think there could be something in it.'

The toddler threw his cup overboard. Ingrid retrieved it, winning a smile.

'Where would you start?' Nick asked.

'Ministry of Agriculture might be worth a try. Remember the droughts in '75 and '76 when people were queuing at standpipes? Imagine being able to make it rain to sort that lot out.'

'But according to this,' he pointed to the faded text on the back of the photograph, 'your Private Dean was talking about the early 1950s. Any experiments couldn't have been much good if we were stuck with a drought twenty years later.'

Ingrid shrugged. 'Maybe, but think of the spread we could run if it's true. Taxpayers money being spent on making it rain, in our climate!'

Nick pushed his empty plate away. 'I suppose Danny could take a look, but it'd be a waste of his time.'

'No!' She didn't mean to react so violently. Danny and Mat were two young career reporters at the paper. They wouldn't be staying twenty-five years; just long enough to get noticed by a national. Of course Nick was going to pass anything controversial onto them. But not this time. This story was hers — it had to be.

Seeds of Doubt is available on
Kindle at £2.99 or £7.99 in paperback.

About the Author

Pamela Vass was drawn to Devon thirty years ago and only afterwards discovered a family tree firmly rooted in Devon soil. Since then it has provided inspiration for her writing. *Seeds of Doubt* grew from suspicions that outside agencies played a part in the 1952 Lynmouth floods, suspicions that provide a gripping background for her novel.

Pamela's career includes several years as Director of *The Whodunnit Company* offering murder mystery events in this country and abroad. Prior to this she was a social worker with Barnardos and two Social Services Departments. This professional experience provides a strong foundation for **Shadow Child,** a gritty and realistic depiction of the challenges faced by a child abandoned by his mother. He never gives up searching for her, not as a child, not as an adult. But the past casts a long shadow and his quest for the truth threatens his very future.

An interest in historical research led to **The Power of 3**, an account of the nineteenth century Devon inventor, Thomas Fowler. Tragically Fowler's ground-breaking work on the principles later embodied in the modern computer was lost, but a combination of the research carried out by Pamela and the expertise of an international team has reinstated him as a significant figure in computing history.

Pamela continues to live and write in Devon, finding inspiration in the unique landscape and the stories it holds.

For more on Pamela Vass and her writing See:
www.boundstonebooks.co.uk